A. GHIZZETTI · A. OSSICINI

QUADRATURE FORMULAE

QUADRATURE FORMULAE

by

A. GHIZZETTI · A. OSSICINI

Professors of Mathematics, Rome University

AKADEMIE-VERLAG · BERLIN

1970

Erschienen im Akademie-Verlag GmbH, 108 Berlin, Leipziger Straße 3–4

Lizenznummer: 202 · 100/551/70
Gesamtherstellung: VEB Druckerei „Thomas Müntzer“, 582 Bad Langensalza
Bestellnummer: 5783 · ES 19 B 4

PREFACE

Quadrature formulae are generally obtained substituting the integral to evaluate $\int_a^b f(x)\,dx$ with the integral $\int_a^b \varphi(x)\,dx$ where $\varphi(x)$ is a function approximating $f(x)$, whose indefinite integral is expressible in elementary functions. The choice of $\varphi(x)$ is performed using interpolation methods. Thus the problem of approximating the integral of $f(x)$ is reduced to the approximation of $f(x)$ itself, that is to a problem not equivalent to the first one. Therefore, it is reasonable to think that it possible to obtain quadrature formulae, without using interpolation methods. In a paper of 1913, G. PEANO (see G. PEANO [1]) first made an attempt in this way and succeeded in obtaining Cavalieri-Simpson's formula, with an integral expression of the remainder, only by means of integration by parts. This method was systematically employed by R. VON MISES, who (in a paper of 1935; see R. VON MISES [1]) showed how it is possible to get every quadrature formulae with the only tool of integration by parts.

In 1935, also J. RADON [1] showed how the integral expression of the remainder can be obtained by means of the Green-Lagrange identity, relative to a linear differential operator and its adjoint. However, these fundamental works remained nearly unknown and still to-day the elementary method suggested by them is not used in every book concerning to quadrature formulae. From 1954, one of us (see A. GHIZZETTI [1], [2], [3], [4]) had the opportunity to reconsider and to probe the study of Radon's procedure, pointing out that it allows to obtain not only the expression of the remainder, but also of the part approximating the integral. This result can be attained in more general conditions than the usual ones, that is making use of Lebesgue integrals for *finite* or *infinite* intervals. In this way a method, which allows to obtain all the known quadrature formulae and many others, is carried out by a systematic application of some simple rules. It is worthwhile to note that Radon's method includes that of von Mises, since the integrations by parts used by von Mises, are already performed in the Green-Lagrange identity, used by Radon.

In 1965, we have reconsidered the study of these methods and their applications (see A. GHIZZETTI [5]; A. OSSICINI [1], [2], [3]; A. GHIZZETTI and A. OSSICINI [1]), availing us of the valued help of some collaborators of ours. In this way we succeeded in collecting all necessary elements to draw up this book, in which we expound the theory of quadrature formulae, only by means of the above mentioned methods.

This book is divided into six chapters. Chapter 6 must be considered independently; it contains exclusively the solutions to all the problems, proposed to the reader in every preceding chapter. Chapter 1 is devoted to the exposition of some preliminary notions about differential linear operators which possess *adjoint operator*. In Chapter 2, as application of the preceding chapter, the general rule for constructing all the quadrature formulae, having a structure analogous to the classical ones, is given and discussed. We call these formulae *elementary* to distinguish them from the *generalized* ones, considered in Chapter 5. These formulae are obtained dividing the integration interval in partial intervals and applying an elementary formula to every one of them. These generalized formulae are just those, which we use for the numerical computation of integrals; on this subject some questions arise about convergence, which are considered in Chapter 5 itself.

Chapters 1, 2, 5 are the most important and this book could be limited to them, since Chapters 3, 4 are exclusively devoted to some applications of the general rule expounded in Chapter 2. They could be considered just an collection of problems. However also Chapters 3, 4 are important, since not only they explain, with concrete examples, the power of the general theory, but in addition they present some very difficult problems, whose solutions requires the knowledge of notions and special techniques, useful in some other cases. For this reason, in Chapter 3 we have collected, for reader's convenience, some elements about certain special functions, which occur in the extensive collection of examples of Chapter 4. These examples have been treated by the same method; what varies from one example to another is the technique of remainder estimate, by means of the integral representation provided by the method.

Since in this book we make use of a unique method, which few Authors have considered, the closely related Bibliography would be not large; therefore we have added the list of the most recent books on quadrature formulae; to these books we refer readers for many practical questions (for example, numerical table about nodes and coefficients of the classical formulae). In some paragraphs, we have put a bibliographic note to point out some papers on which we have founded the draft of the paragraph itself. We are grateful to "Birkhäuser Verlag" which has received this books in one of its valuable collections.

Rome, October, 1969 A. Ghizzetti · A. Ossicini

REMARK

Every chapter is divided into paragraphs, which are denoted by a symbol as § 3.4 (to indicate the fourth paragraph of Chapter 3). In every paragraph, formulas are indicated by three numbers into brackets; for instance (3.4.7) means the 7th formula of § 3.4. For theorems, an analogous notation is used, without brackets, and having the 3th number Roman; for instance, theor. 3.4.II means the second theorem of § 3.4. In the end of every chapter 1, 2, 3, 4, 5 there is a paragraph concerning problems proposed to readers; their solutions are collected in Chapter 6. These problems are denoted as § 1.5., Problem 2, but of course the corresponding solution of Chapter 6 must be considered.

TABLE OF CONTENTS

CHAPTER 1

Additional results on linear differential equations

1.1. Linear differential operators and their adjoints. In this chapter we shall consider exclusively real-valued functions. If $[a, b]$ is a finite interval, we shall call $L[a, b]$ the class of functions $f(x)$ Lebesgue-integrable (summable) in $[a, b]$ and $A\ C^k[a, b]$ the class of functions $f(x)$ whose k-th derivative $f^{(k)}(x)$ is absolutely continuous in $[a, b]$ $(k = 0, 1, 2, \ldots)$.[1]) It is known that if $f(x) \in A\ C^k[a, b]$ the derivative $f^{(k+1)}(x)$ exists almost everywhere in $[a, b]$ and that $f^{(k+1)}(x) \in L[a, b]$.

Let us consider in $[a, b]$ a *linear differential operator of order* n $(n = 1, 2, 3, \ldots)$:

$$E = \sum_{k=0}^{n} a_k(x) \frac{d^{n-k}}{dx^{n-k}} \tag{1.1.1}$$

with the following conditions on the coefficients $a_k(x)$

$$a_0(x) = 1; \quad a_k(x) \in A\ C^{n-k-1}[a, b], \quad (k = 1, 2, \ldots, n-1); \quad a_n(x) \in L[a, b). \tag{1.1.2}$$

The operator E can be applied to the functions $u(x) \in A\ C^{n-1}[a, b]$, obtaining the function (defined almost everywhere):

$$E[u(x)] = E(u) = \sum_{k=0}^{n} a_k(x)\, u^{(n-k)}(x) \in L[a, b]. \tag{1.1.3}$$

In the following we shall have occasion to associate with the operator E the *reduced operators*

$$E_r = \sum_{k=0}^{r} a_k(x) \frac{d^{r-k}}{dx^{r-k}}, \quad (r = 0, 1, \ldots, n-1).^{2)} \tag{1.1.4}$$

By virtue of the hypotheses (1.1.2) we can consider the so-called *adjoint operator* E^* of the operator E, defined by

$$E^* = \sum_{k=0}^{n} (-1)^{n-k} \frac{d^{n-k}}{dx^{n-k}} a_k(x). \tag{1.1.5}$$

The adjoint operator can be also written

$$E^* = \sum_{k=0}^{n} a_k^*(x) \frac{d^{n-k}}{dx^{n-k}} \tag{1.1.6}$$

[1]) For $k = 0$ we have the class $A\ C^0[a, b]$ of absolutely continuous functions; this class is also denoted by $A\ C[a, b]$.

[2]) For $r = n$ we obtain $E_n = E$.

and an easy computation shows that

$$a_k^*(x) = \sum_{i=0}^{k} (-1)^{n-i} \binom{n-i}{k-i} a_i^{(k-i)}(x), \quad (k = 0, 1, \ldots, n). \quad (1.1.7)$$

From (1.1.2) and (1.1.7) easily follow

$$a_0^*(x) = (-1)^n; \quad a_k^*(x) \in A\, C^{n-k-1}[a, b], \ (k = 1, 2, \ldots, n-1); \quad a_n^*(x) \in L[a, b], \quad (1.1.8)$$

analogous to (1.1.2), so that the operator E^* can be applied to the functions $v(x) \in A\ C^{n-1}[a, b]$ obtaining a function (defined almost everywhere):

$$E^*[v(x)] = E^*(v) = \sum_{k=0}^{n} (-1)^{n-k} [a_k(x)\, v(x)]^{(n-k)} \in L[a, b]. \quad (1.1.9)$$

From (1.1.8) it follows that we can consider the operator E^{**}, adjoint to E^*, but we do not obtain a new operator, since the reader will easily prove that $E^{**} = E$.

We shall also have the occasion to consider the adjoint operator E_r^* of the reduced operators E_r, for which we obviously have

$$E_r^*(v) = \sum_{k=0}^{r} (-1)^{r-k} [a_k(x)\, v(x)]^{(r-k)}, \quad (r = 0, 1, \ldots, n-1). \quad (1.1.10)$$

Note also that there exists the recurrence formula

$$E_r^*(v) = -\frac{d}{dx} E_{r-1}^*(v) + a_r(x)\, v(x), \quad (r = 1, 2, \ldots, n).^{1)} \quad (1.1.11)$$

1.2. The Green-Lagrange identity. We shall begin by proving that, if $u(x)$, $v(x) \in A\ C^{k-1}[a, b]$, $(k = 1, 2, 3, \ldots)$ the relation

$$v(x)\, u^{(k)}(x) - (-1)^k u(x)\, v^{(k)}(x) = \frac{d}{dx} \sum_{i=0}^{k-1} (-1)^{k-i-1} u^{(i)}(x)\, v^{(k-i-1)}(x) \quad (1.2.1)$$

holds almost everywhere in $[a, b]$.

We prove it by induction, observing first of all that for $k = 1$, (1.2.1) is evident since it becomes $v\, u' + u\, v' = \frac{d}{dx}(u\, v)$.

To pass from k to $k + 1$, it is thus sufficient to write

$$v\, u^{(k+1)} - (-1)^{k+1} u\, v^{(k+1)} = \frac{d}{dx}[(-1)^k u\, v^{(k)}] + \left[v \left(\frac{du}{dx}\right)^{(k)} - (-1)^k \frac{du}{dx} v^{(k)}\right]$$

$$= \frac{d}{dx}\left[(-1)^k u\, v^{(k)} + \sum_{i=0}^{k-1} (-1)^{k-i-1} \left(\frac{du}{dx}\right)^{(i)} v^{(k-i-1)}\right]$$

$$= \frac{d}{dx}\left[(-1)^k u\, v^{(k)} + \sum_{i=1}^{k} (-1)^{k-i} u^{(i)}\, v^{(k-i)}\right] = \frac{d}{dx} \sum_{i=0}^{k} (-1)^{k-i} u^{(i)}\, v^{(k-i)}.$$

1) For $r = n$, remember that $E_n^* = E^*$.

Now we can prove the Green-Lagrange identity, relative to a linear differential operator E of order n of the type (1.1.1) under the hypotheses (1.1.2) and to its adjoint E^* [see (1.1.5)]. The Green-Lagrange identity states that for every pair of functions $u(x)$, $v(x) \in A\ C^{n-1}[a, b]$, the following equation holds almost everywhere in $[a, b]$:

$$v(x)\, E[u(x)] - u(x)\, E^*[v(x)] = \frac{d}{dx} \sum_{k=0}^{n-1} u^{(k)}(x)\, E^*_{n-k-1}[v(x)]\,, \qquad (1.2.2)$$

where the operators E^*_{n-k-1} are defined by (1.1.10). We have

$$v\, E(u) - u\, E^*(v) = v \sum_{i=0}^{n} a_i\, u^{(n-i)} - u \sum_{i=0}^{n} (-1)^{n-i} (a_i\, v)^{(n-i)}$$
$$= \sum_{i=0}^{n-1} [a_i\, v \cdot u^{(n-i)} - (-1)^{n-i}\, u \cdot (a_i\, v)^{(n-i)}]\,,$$

and applying (1.2.1) to the functions $u \in A\ C^{n-1}[a, b]$, $a_i\, v \in A\ C^{n-i-1}[a, b]$

$$v\, E(u) - u\, E^*(v) = \sum_{i=0}^{n-1} \frac{d}{dx} \sum_{k=0}^{n-i-1} (-1)^{n-i-k-1}\, u^{(k)}\, (a_i\, v)^{(n-i-k-1)}$$
$$= \frac{d}{dx} \sum_{k=0}^{n-1} u^{(k)} \sum_{i=0}^{n-k-1} (-1)^{n-k-1-i}\, (a_i\, v)^{(n-k-1-i)}\,;$$

from which, using (1.1.10), (1.2.2) follows.

1.3. Relations between the solutions of a linear differential equation and those of the adjoint equation. Let us use the symbols E, E^* with the meaning already given in (1.1.1), (1.1.2) and (1.1.5). Let us consider in $[a, b]$ the non-homogeneous linear differential equation

$$E(u) = f(x) \qquad (1.3.1)$$

with $f(x) \in L[a, b]$. Let $[u_1(x), u_2(x), \ldots, u_n(x)]$, where $u_i \in A\ C^{n-1}[a, b]$, be a fundamental system of solutions of the homogeneous equation $E(u) = 0$, and let $U(x) \neq 0$ their Wronskian. It is known that the general solution of (1.3.1) is expressed by

$$u(x) = \sum_{i=1}^{n} \alpha_i\, u_i(x) + \int_a^x K(x, \xi)\, f(\xi)\, d\xi\,, \qquad (1.3.2)$$

where $\alpha_1, \alpha_2, \ldots, \alpha_n$ are arbitrary constants and $K(x, \xi)$ is the so-called *Cauchy resolvent kernel*, which is (as function of x) the particular solution of the homogeneous equation $E(u) = 0$ which satisfies, at the point ξ, the following initial conditions:

$$\left[\frac{\partial^h}{\partial x^h} K(x, \xi)\right]_{x=\xi} = \delta_{h,\, n-1}\,, \quad (h = 0, 1, \ldots, n-1)\,.^{1)} \qquad (1.3.3)$$

[1]) δ_{rs} is the Kronecker delta $= 0$ (if $r \neq s$)
$= 1$ (if $r = s$).

It may be written

$$K(x, \xi) = \sum_{i=1}^{n} u_i(x)\, v_i(\xi)\,, \tag{1.3.4}$$

where the functions $v_1(x), v_2(x), \ldots, v_n(x)$ are given by the system of n linear equations

$$\sum_{i=1}^{n} u_i^{(h)}(x)\, v_i(x) = \delta_{h,\, n-1}\,, \quad (h = 0, 1, \ldots, n-1)\,, \tag{1.3.5}$$

whose determinant is $U(x)$.[1]) Therefore (1.3.2) may also be written

$$u(x) = \sum_{i=1}^{n} u_i(x) \left[\alpha_i + \int_a^x v_i(\xi) f(\xi)\, d\xi\right]. \tag{1.3.6}$$

As to the derivatives of $u(x)$, it is well known that from (1.3.2) and (1.3.6) follows

$$u^{(h)}(x) = \sum_{i=1}^{n} \alpha_i\, u_i^{(h)}(x) + \int_a^x \frac{\partial^h K(x, \xi)}{\partial x^h} f(\xi)\, d\xi$$

$$= \sum_{i=1}^{n} u_i^{(h)}(x) \left[\alpha_i + \int_a^x v_i(\xi) f(\xi)\, d\xi\right], \quad (h = 0, 1, \ldots, n-1)\,. \tag{1.3.7}$$

We observe another property of the functions $v_i(x)$ introduced by (1.3.5); we have almost everywhere

$$\sum_{i=1}^{n} u_i^{(n)} v_i = \sum_{i=1}^{n}\left[-\sum_{k=1}^{n} a_k\, u_i^{(n-k)}\right] v_i = -\sum_{k=1}^{n} a_k \sum_{i=1}^{n} u_i^{(n-k)}\, v_i = -\sum_{k=1}^{n} a_k\, \delta_{n-k,\, n-1}\,,$$

that is

$$\sum_{i=1}^{n} u_i^{(n)}(x)\, v_i(x) = -\, a_1(x)\,. \tag{1.3.8}$$

The relations (1.3.5), (1.3.8) may be generalized in the way shown by the following theorem:

Theorem 1.3.I. *The functions $v_i(x)$ defined by the system (1.3.5) belong to the class $A\, C^{n-1}[a, b]$ and satisfy the following relations*

$$\sum_{i=1}^{n} u_i^{(h)}(x)\, E_k^*[v_i(x)] = \delta_{h,\, n-k-1}\,, \quad (h, k = 0, 1, \ldots, n-1)\,, \tag{1.3.9}$$

$$\sum_{i=1}^{n} u_i^{(n)}(x)\, E_k^*[v_i(x)] = -\, a_{k+1}(x)\,, \quad (k = 0, 1, \ldots, n-1)\,,^{2)} \tag{1.3.10}$$

where E_k^* are the operators defined by (1.1.10).

Proof. Since $E_0^*(v_i) = v_i$, for $k = 0$ the equation (1.3.9) reduce to (1.3.5) (which assures $v_i \in A\, C$) and (1.3.10) reduces to (1.3.8). We proceed by

[1]) From (1.3.5) and from the fact that $u_i(x) \in A\, C^{n-1}[a, b]$ it follows immediately that $v_i(x) \in A\, C[a, b]$; but we shall prove shortly that $v_i(x) \in A\, C^{n-1}[a, b]$ (see theorem 1.3.I).

[2]) (1.3.10) holds almost everywhere.

induction to assume that for a certain k (with $0 \leqq k < n - 1$) the equation (1.3.9) hold, that is $E_k^*(v_i) \in A\,C$ or $v_i \in A\,C^k$. We prove that as a consequence (1.3.10) holds, with the same k and (1.3.9) with $k+1$ instead of k; thus $v_i \in A\,C^{k+1}$. We may prove (1.3.10) immediately by observing that

$$\sum_{i=1}^{n} u_i^{(n)} E_k^*(v_i) = \sum_{i=1}^{n}\left[-\sum_{h=1}^{n} a_h u_i^{(n-h)}\right] E_k^*(v_i) = -\sum_{h=1}^{n} a_h \sum_{i=1}^{n} u_i^{(n-h)} E_k^*(v_i)$$

$$= -\sum_{h=1}^{n} a_h \delta_{n-h,\,n-k-1} = -a_{k+1}\,.$$

To prove the other assertion, we start from (1.3.9) and differentiate both members; there results almost everywhere

$$\sum_{i=1}^{n} u_i^{(h+1)} E_k^*(v_i) + \sum_{i=1}^{n} u_i^{(h)} \frac{d}{dx} E_k^*(v_i) = 0\,, \quad (h = 0, 1, \ldots, n-1)\,. \tag{1.3.11}$$

The value of the first term is $\delta_{h+1,\,n-k-1}$ if $h = 0, 1, \ldots, n-2$ [from (1.3.9)] and $-a_{k+1}$ if $h = n-1$ [from (1.3.10)]. For the second term we have from (1.1.11)

$$\sum_{i=1}^{n} u_i^{(h)} \frac{d}{dx} E_k^*(v_i) = \sum_{i=1}^{n} u_i^{(h)} [-E_{k+1}^*(v_i) + a_{k+1} v_i]$$

$$= -\sum_{i=1}^{n} u_i^{(h)} E_{k+1}^*(v_i) + a_{k+1} \delta_{h,\,n-1}\,.$$

With this, (1.3.11) yields

$$\left.\begin{aligned} &\delta_{h+1,\,n-k-1} - \sum_{i=1}^{n} u_i^{(h)} E_{k+1}^*(v_i) + 0 = 0\,, \quad (h = 0, 1, \ldots, n-2)\,,\\ &-a_{k+1} - \sum_{i=1}^{n} u_i^{(n-1)} E_{k+1}^*(v_i) + a_{k+1} = 0 \end{aligned}\right\} \tag{1.3.12}$$

which may be gathered in

$$\sum_{i=1}^{n} u_i^{(h)} E_{k+1}^*(v_i) = \delta_{h,\,n-k-2}\,, \quad (h = 0, 1, \ldots, n-1)\,, \tag{1.3.13}$$

which is precisely (1.3.9) with $k+1$ instead of k, q.e.d.

We now prove another theorem:

Theorem 1.3.II. *The functions* $[v_1(x), v_2(x), \ldots, v_n(x)]$ *defined by the system (1.3.5) form a fundamental system of solutions of the homogeneous adjoint equation*

$$E^*(v) = 0 \tag{1.3.14}$$

and for their Wronskian $V(x)$ *we have*

$$V(x) = \frac{1}{U(x)}\,. \tag{1.3.15}$$

The general solution of the non-homogeneous adjoint equation

$$E^*(v) = g(x)\,, \qquad \big(g(x) \in L[a, b]\big)\,, \tag{1.3.16}$$

is given by

$$v(x) = \sum_{i=1}^{n} \beta_i\, v_i(x) + \int_a^x K^*(x, \xi)\, g(\xi)\, d\xi\,, \tag{1.3.17}$$

where $\beta_1, \beta_2, \ldots, \beta_n$ *are arbitrary constants, while for the resolvent kernel* $K^*(x, \xi)$ *we have*

$$K^*(x, \xi) = -K(\xi, x) = -\sum_{i=1}^{n} v_i(x)\, u_i(\xi)\,. \tag{1.3.18}$$

Therefore (1.3.17) *may also be written*

$$v(x) = \sum_{i=1}^{n} v_i(x)\left[\beta_i - \int_a^x u_i(\xi)\, g(\xi)\, d\xi\right]. \tag{1.3.19}$$

Proof. We begin by proving that $E^*(v_i) = 0$ $(i = 1, 2, \ldots, n)$. Starting from (1.3.9) written with $k = n - 1$, we may proceed in the same fashion as in (1.3.11), (1.3.12), (1.3.13) (with $k = n - 1$, $E_n^* = E^*$) thus obtaining

$$\sum_{i=1}^{n} u_i^{(h)}\, E^*(v_i) = 0\,, \qquad (h = 0, 1, \ldots, n-1)\,,$$

which, together with $U(x) \neq 0$, give $E^*(v_i) = 0$.

To prove (1.3.15) we observe that from (1.1.10) it easily follows that the derivative $v^{(r)}(x)$, $(r = 0, 1, \ldots, n-1)$ of a $v(\ ) \in A\, C^{n-1}[a, b]$ may be expressed by means of $E_r^*(v), E_{r-1}^*(v), \ldots, E_0^*(v)$ with a formula of the type

$$v^{(r)}(x) = (-1)^r\, E_r^*[v(x)] + \sum_{s=1}^{r} c_{rs}(x)\, E_{r-s}^*[v(x)]\,. \tag{1.3.20}$$

Thus we may write

$$V(x) = \begin{vmatrix} v_1 & v_2 & \ldots & v_n \\ v_1' & v_2' & \ldots & v_n' \\ v_1'' & v_2'' & \ldots & v_n'' \\ \cdot & \cdot & \cdot & \cdot \\ v_1^{(n-1)} & v_2^{(n-1)} & \ldots & v_n^{(n-1)} \end{vmatrix}$$

$$= \begin{vmatrix} E_0^*(v_1) & E_0^*(v_2) & \ldots & E_0^*(v_n) \\ -E_1^*(v_1) & -E_1^*(v_2) & \ldots & -E_1^*(v_n) \\ E_2^*(v_1) & E_2^*(v_2) & \ldots & E_2^*(v_n) \\ \cdot & \cdot & \cdot & \cdot \\ (-1)^{n-1} E_{n-1}^*(v_1) & (-1)^{n-1} E_{n-1}^*(v_2) & \ldots & (-1)^{n-1} E_{n-1}^*(v_n) \end{vmatrix}$$

$$= (-1)^{\binom{n}{2}} \begin{vmatrix} E_0^*(v_1) & E_0^*(v_2) & \ldots & E_0^*(v_n) \\ E_1^*(v_1) & E_1^*(v_2) & \ldots & E_1^*(v_n) \\ \cdot & \cdot & \cdot & \cdot \\ E_{n-1}^*(v_1) & E_{n-1}^*(v_2) & \ldots & E_{n-1}^*(v_n) \end{vmatrix}$$

and then, computing $V(x)\ U(x)$ as the determinant of the product of this matrix by

$$\begin{pmatrix} u_1 & u_1' & \dots & u_1^{(n-1)} \\ u_2 & u_2' & \dots & u_2^{(n-1)} \\ \vdots & \vdots & & \vdots \\ u_n & u_n' & \dots & u_n^{(n-1)} \end{pmatrix}$$

we get, by virtue of (1.3.9)

$$V(x)\,U(x) = (-1)^{\binom{n}{2}} \begin{vmatrix} \delta_{0,n-1} & \delta_{1,n-1} & \dots & \delta_{n-1,n-1} \\ \delta_{0,n-2} & \delta_{1,n-2} & \dots & \delta_{n-1,n-2} \\ \dots & \dots & \dots & \dots \\ \delta_{0,0} & \delta_{1,0} & \dots & \delta_{n-1,0} \end{vmatrix} = (-1)^{\binom{n}{2}} \begin{vmatrix} 0\,0 \dots 0\,1 \\ 0\,0 \dots 1\,0 \\ \dots \\ 0\,1 \dots 0\,0 \\ 1\,0 \dots 0\,0 \end{vmatrix} = 1 .$$

Finally we prove (1.3.18). Considering that in the first member of the differential equation (1.3.16) the coefficient of $v^{(n)}$ is not equal to 1, but to $(-1)^n$ [see (1.1.8)], we may assert that the resolvent kernel $K^*(x, \xi)$ is given by $\sum_{i=1}^{n} v_i(x)\, w_i(\xi)$, where the functions $w_1(x), w_2(x), \dots, w_n(x)$ are defined by the system of linear equations

$$\sum_{i=1}^{n} v_i^{(h)}\, w_i = (-1)^n\, \delta_{h,\,n-1}\,, \qquad (h = 0, 1, \dots, n-1)\,.$$

By application of equation (1.3.20), this system may also be written

$$\sum_{i=1}^{n} \left[(-1)^h\, E_h^*(v_i) + \sum_{s=1}^{h} c_{hs}\, E_{h-s}^*(v_i)\right] w_i = (-1)^n\, \delta_{h,\,n-1}\,,$$
$$(h = 0, 1, \dots, n-1)\,,$$

or also, by subtracting from every equation an obvious linear combination of the preceding ones,

$$\sum_{i=1}^{n} E_h^*(v_i)\, w_i = (-1)^{n-h}\, \delta_{h,\,n-1}\,, \qquad (h = 0, 1, \dots, n-1)\,.$$

Finally another transformation leads to

$$\sum_{i=1}^{n} E_h^*(v_i)\,(-\,w_i) = \delta_{h,\,n-1}\,, \qquad (h = 0, 1, \dots, n-1)\,. \qquad (1.3.21)$$

But from (1,3.9) we have

$$\sum_{i=1}^{n} E_h^*(v_i)\, u_i = \delta_{0,\,n-h-1} = \delta_{h,\,n-1}\,, \qquad (h = 0, 1, \dots, n-1)\,, \qquad (1.3.22)$$

and since the determinant of $E_h^*(v_i)$ is $(-1)^{\binom{n}{2}}\, V(x) \neq 0$, the comparison of (1.3.21) with (1.3.22) leads to $w_i = -\,u_i$. Equation (1.3.18) follows immediately.

To complete this theorem, we note that from (1.3.17) and (1.3.19) we get

$$E^*_{n-h-1}[v(x)] = \sum_{i=1}^{n} \beta_i \, E^*_{n-h-1}[v_i(x)] + \int_a^x E^*_{n-h-1}[K^*(x, \xi)] \, g(\xi) \, d\xi$$

$$= \sum_{i=1}^{n} E^*_{n-h-1}[v_i(x)] \left[\beta_i - \int_a^x u_i(\xi) \, g(\xi) \, d\xi\right]. \qquad (1.3.23)$$

A fundamental system $(u_1, u_2, \ldots, u_n)$ of solutions of $E(u) = 0$ and a fundamental system $(v_1, v_2, \ldots, v_n)$ of solutions of $E^*(v) = 0$ will be called *associated* if the latter can be obtained from the first using (1.3.5).

The following theorem will also be useful:

Theorem 1.3.III. *With the meaning previously given to $u_i(x)$ and $v_i(x)$ the formula*

$$\sum_{k=0}^{n-1} u_i^{(k)}(x) \, E^*_{n-k-1} \, [v_j(x)] = \delta_{ij} , \qquad (i, j = 1, 2, \ldots, n) \qquad (1.3.24)$$

holds.

Proof. Let us denote by λ_{ij} the first member of (1.3.24). Holding the index i fixed, for every $h = 0, 1, \ldots, n-1$ we may write

$$\sum_{j=1}^{n} u_j^{(h)} \lambda_{ij} = \sum_{j=1}^{n} u_j^{(h)} \sum_{k=0}^{n-1} u_i^{(k)} \, E^*_{n-k-1}(v_j) = \sum_{k=0}^{n-1} u_i^{(k)} \sum_{j=1}^{n} u_j^{(h)} \, E^*_{n-k-1}(v_j)$$

and thus by (1.3.9)

$$\sum_{j=1}^{n} u_j^{(h)} \lambda_{ij} = \sum_{k=0}^{n-1} u_i^{(k)} \, \delta_{h,k} = u_i^{(h)} , \qquad (h = 0, 1, \ldots, n-1) .$$

This system of n linear equations in n unknowns $\lambda_{i1}, \lambda_{i2}, \ldots, \lambda_{in}$ in virtue of the fact that $U(x) \neq 0$, has only the solution $\lambda_{ij} = \delta_{ij}$ which establishes (1.3.24), q.e.d.

1.4. Linear differential equations considered in infinite intervals.[1] The statements made in § 1,1, 1.2 and 1.3 pertain to the case in which the independent variable x varies in a finite interval $[a, b]$. With slight modifications they hold also when x varies in an infinite interval, for instance $[a, +\infty)$.[2] It suffices to substitute for the class $L[a, b]$ the class $L_{\text{loc}}\,[a, +\infty)$ of *locally summable* functions in $[a, +\infty)$ and for the class $AC^k[a, b]$ the class $AC^k_{\text{loc}}\,[a, +\infty)$ of functions whose *k-th derivative is locally absolutely continuous* in $[a, +\infty)$.[3] With only these modifications all the formulae and theorems of § 1.1, 1.2, 1.3 remain valid. We shall now give some other basic ideas essentially connected with the infinite interval.

[1]) See A. Ghizzetti [2], [3], [4].

[2]) Here we only refer to the case of $[a, +\infty)$; similar considerations hold for the interval $(-\infty, +\infty)$, and we shall summarily state them at the end of the paragraph.

[3]) A function $f(x)$ is called *locally summable* (or *locally absolutely continuous*) in $[a, +\infty)$ when it is summable (or absolutely continuous) in every finite interval contained in $[a, +\infty)$.

Writing $f(x) \in I\,[a, +\infty)$, that is, $f(x)$ is *integrable* in $[a, +\infty)$, we shall mean that $f(x)$ is locally summable in $[a, +\infty)$ and moreover that the integral $\int_a^{+\infty} f(x)\,dx$ is *convergent.*[1]) In a particular case it may happen that this integral be *absolutely convergent*; we note that this is the same as saying that $f(x)$ is Lebesgue-integrable (summable) on the interval $[a, +\infty)$, that is $f(x) \in L\,[a, +\infty)$. Let us now consider again the differential equations (1.3.1) and (1.3.16), namely

$$E(u) = f(x) \qquad (1.4.1)$$

$$E^*(v) = g(x) \qquad (1.4.2)$$

under the new hypotheses $f(x), g(x) \in L_{\text{loc}}[a, +\infty)$.

From § 1.3 we know that having introduced two *associated systems*, $(u_1, \ldots, u_n)$, $(v_1, \ldots, v_n)$ of solutions of the homogeneous equations $E(u) = 0$, $E^*(v) = 0$, we can derive from (1.4.1) and (1.4.2) the equations [see (1.3.7) and (1.3.23)]:

$$u^{(h)}(x) = \sum_{i=1}^{n} u_i^{(h)}(x)\left[\alpha_i + \int_a^x f(\xi)\, v_i(\xi)\, d\xi\right], \quad (h = 0, 1, \ldots, n-1)\,, \qquad (1.4.3)$$

$$E^*_{n-h-1}[v(x)] = \sum_{i=1}^{n} E^*_{n-h-1}[v_i(x)]\left[\beta_i - \int_a^x g(\xi)\, u_i(\xi)\, d\xi\right],$$
$$(h = 0, 1, \ldots, n-1)\,, \qquad (1.4.4)$$

where α_i, β_i are arbitrary constants.

Now let us assume that *the term denoted by $g(x)$ satisfies the additional hypotheses*

$$g(x)\, u_i(x) \in I\,[a, +\infty)\,, \quad (i = 1, 2, \ldots, n)\,. \qquad (1.4.5)$$

We choose the arbitrary constants β_i so that $\lim_{x\to+\infty}\left[\beta_i - \int_a^x g\, u_i\, d\xi\right] = 0$; it is evident that we must put

$$\beta_i = \int_a^{+\infty} g(\xi)\, u_i(\xi)\, d\xi\,, \quad (i = 1, 2, \ldots, n)\,. \qquad (1.4.6)$$

In this way we specify a certain particular solution of (1.4.2) that we shall denote by $V(x)$ and for which (1.4.4) becomes

$$E^*_{n-h-1}[V(x)] = \sum_{i=1}^{n} E^*_{n-h-1}\,[v_i(x)] \int_x^{+\infty} g(\xi)\, u_i(\xi)\, d\xi\,,$$
$$(h = 0, 1, \ldots, n-1)\,, \qquad (1.4.7)$$

[1]) This means that the $\lim_{x\to+\infty} \int_a^x f(x)\,dx$ is finite. In the case of the interval $(-\infty, +\infty)$ both the limits $\lim_{x\to-\infty} \int_x^0 f(x)\,dx$ and $\lim_{x\to+\infty} \int_0^x f(x)\,dx$ must be finite.

and in particular for $h = n - 1$

$$V(x) = \sum_{i=1}^{n} v_i(x) \int_x^{+\infty} g(\xi)\, u_i(\xi)\, d\xi \tag{1.4.8}$$

that is, recalling (1.3.18)

$$V(x) = \int_x^{+\infty} K(\xi, x)\, g(\xi)\, d\xi\,. \tag{1.4.9}$$

This particular solution $V(x)$ will be denoted as that particular solution of (1.4.2) *which satisfies the initial conditions null at the point* $+\infty$. This sentence is not to be taken literally; in fact generally it is *not* true that

$$\lim_{x\to+\infty} V^{(h)}(x) = 0\,, \quad (h = 0, 1, \ldots, n-1)\,.^{1)} \tag{1.4.10}$$

However, we shall use it for the following three reasons:

1°) if it is true that in general (1.4.10) do not hold, *although n linear independent combinations of* $V^{(h)}(x)$ *do exist* (with coefficient functions of x) *which tend to zero for* $x \to +\infty$, we have in fact

$$\sum_{h=0}^{n-1} u_i^{(h)}(x)\, E^*_{n-h-1}[V(x)] = \text{[using (1.4.7)]}$$

$$= \sum_{h=0}^{n-1} u_i^{(h)}(x) \sum_{j=1}^{n} E^*_{n-h-1}[v_j(x)] \int_x^{+\infty} g(\xi)\, u_j(\xi)\, d\xi$$

$$= \sum_{j=1}^{n} \int_x^{+\infty} g(\xi)\, u_j(\xi)\, d\xi \sum_{h=0}^{n-1} u_i^{(h)}(x)\, E^*_{n-h-1}[v_j(x)] = \text{[using (1.3.24)]}$$

$$= \sum_{j=1}^{n} \int_x^{+\infty} g(\xi)\, u_j(\xi)\, d\xi \cdot \delta_{ij} = \int_x^{+\infty} g(\xi)\, u_i(\xi)\, d\xi$$

and therefore

$$\lim_{x\to+\infty} \sum_{h=0}^{n-1} u_i^{(h)}(x)\, E^*_{n-h-1}[V(x)] = 0\,, \quad (i = 1, 2, \ldots, n);^{2)} \tag{1.4.11}$$

[1]) For instance, having assumed $n = 1$, $E(u) = u' + \frac{1}{x} u$, $E^*(v) = -v' + \frac{1}{x} v$, $g(x) = \frac{1}{x}$ in the interval $[1, +\infty)$ we have $u_1(x) = \frac{1}{x}$, $v_1(x) = x$. The hypothesis $g\, u_1 \in I\,[1, +\infty)$ is satisfied and it turns out that

$$V(x) = x \int_x^{+\infty} \frac{1}{\xi} \cdot \frac{1}{\xi}\, d\xi = 1\,.$$

[2]) In the example of the preceding footnote, (1.4.11) becomes

$$u_1(x)\, V(x) = \frac{1}{x} \cdot 1 \to 0 \quad (\text{for } x \to +\infty)\,.$$

2°) the expression (1.4.9) of $V(x)$ is similar to the expression $\int_x^{x_0} K(\xi, x)\, g(\xi)\, d\xi$ of the solution of (1.4.2) which satisfies the initial conditions null at the point x_0; [1])

3°) having introduced the above mentioned sentence, in Chapter 2, we shall have the advantage to be able to use the same language for the quadrature formulae on finite intervals and for those on infinite intervals.

In connection with this solution $V(x)$ of (1.4.2) and with *any solution* $u(x)$ *of* (*1.4.1.*), we consider the following function

$$w(x) = \sum_{h=0}^{n-1} u^{(h)}(x)\, E^*_{n-h-1}[V(x)]\,. \text{ [2])} \tag{1.4.12}$$

In the next chapter we shall see that, in the theory of quadrature formulae on infinite intervals, it is interesting to discover under what other conditions [besides (1.4.5)] we may assert that

$$\lim_{x\to+\infty} w(x) = 0\,. \tag{1.4.13}$$

Thus we now give the following theorem:

Theorem 1.4.I. *If the given function* $g(x)$ *in equation (1.4.2) verifies the following hypotheses [a particular case of (1.4.5)]*

$$g(x)\, u_i(x) \in L\,[a, +\infty)\,, \quad (i = 1, 2, \ldots, n) \tag{1.4.14}$$

and if the given function $f(x)$ *in equation (1.4.1) satisfies these other hypotheses*

$$f(x)\, v_i(x) \int_x^{+\infty} |g(\xi)\, u_i(\xi)|\, d\xi \in L\,[a, +\infty)\,, \quad (i = 1, 2, \ldots, n)\,, \tag{1.4.15}$$

then, for every solution $u(x)$ *of (1.4.1) and for that particular solution* $V(x)$ *of (1.4.2) which satisfies the initial conditions null at the point* $+\infty$, *we have*

$$g(x)\, u(x) \in L\,[a, +\infty)\,, \tag{1.4.16}$$

$$f(x)\, V(x) \in L\,[a, +\infty)\,, \tag{1.4.17}$$

$$\lim_{x\to+\infty} w(x) = \lim_{x\to+\infty} \sum_{h=0}^{n-1} u^{(h)}(x)\, E^*_{n-h-1}[V(x)] = 0\,. \tag{1.4.18}$$

Proof. Equation (1.4.15) says that

$$\int_a^{+\infty} |f(x)\, v_i(x)|\, dx \int_x^{+\infty} |g(\xi)\, u_i(\xi)|\, d\xi < +\infty\,,$$

whence, by a well-known theorem of Tonelli, the function $f(x)\, v_i(x)\, g(\xi)\, u_i(\xi)$

[1]) If we write (1.4.10) and (1.4.11) with $\lim_{x\to x_0}$, they are equivalent; on the contrary they are not equivalent in the case of the point $+\infty$.

[2]) This function is the same which occurs in the right member of the Green-Lagrange identity (1.2.2).

of the two variables x and ξ is summable in the unbounded domain ($\xi \geqq x \geqq a$) of the (x, ξ) plane. Then applying the Fubini theorem we may write

$$\int_a^{+\infty} f(x)\, v_i(x)\, dx \int_x^{+\infty} g(\xi)\, u_i(\xi)\, d\xi = \int_a^{+\infty} g(\xi)\, u_i(\xi)\, d\xi \int_a^{\xi} f(x)\, v_i(x)\, dx$$

and also, interchanging the variables x and ξ in the second member:

$$\int_a^{+\infty} f(x)\, v_i(x)\, dx \int_x^{+\infty} g(\xi)\, u_i(\xi)\, d\xi = \int_a^{+\infty} g(x)\, u_i(x)\, dx \int_a^{x} f(\xi)\, v_i(\xi)\, d\xi\,. \quad (1.4.19)$$

Thus we may say that from a hypothesis given above we draw not only the obvious conclusion

$$f(x)\, v_i(x) \int_x^{+\infty} g(\xi)\, u_i(\xi)\, d\xi \in L\,[a, +\infty)\,, \qquad (i = 1, 2, \ldots, n) \quad (1.4.20)$$

but also

$$g(x)\, u_i(x) \int_a^{x} f(\xi)\, v_i(\xi)\, d\xi \in L\,[a, +\infty)\,, \qquad (i = 1, 2, \ldots, n) \quad (1.4.21)$$

showing the validity of (1.4.19).

Once that is done, we observe that using (1.4.3), with $h = 0$, we obtain

$$g(x)\, u(x) = \sum_{i=1}^{n} \left[\alpha_i\, g(x)\, u_i(x) + g(x)\, u_i(x) \int_a^{x} f(\xi)\, v_i(\xi)\, d\xi\right],$$

which, together with (1.4.14) and (1.4.21) make (1.4.16) hold and establishes the validity of the formula

$$\int_a^{+\infty} g(x)\, u(x)\, dx = \sum_{i=1}^{n} \left[\alpha_i \int_a^{+\infty} g(x)\, u_i(x)\, dx + \int_a^{+\infty} g(x)\, u_i(x)\, dx \int_a^{x} f(\xi)\, v_i(\xi)\, d\xi]\right]. \quad (1.4.22)$$

From (1.4.8) we get

$$f(x)\, V(x) = \sum_{i=1}^{n} f(x)\, v_i(x) \int_x^{+\infty} g(\xi)\, u_i(\xi)\, d\xi$$

which, together with (1.4.20), proves (1.4.17) and the existence of

$$\int_a^{+\infty} f(x)\, V(x)\, dx = \sum_{i=1}^{n} \int_a^{+\infty} f(x)\, v_i(x)\, dx \int_x^{+\infty} g(\xi)\, u_i(\xi)\, d\xi\,. \quad (1.4.23)$$

Now we observe that the Green-Lagrange identity (1.2.2) gives

$$V\, E(u) - u\, E^*(V) = \frac{d}{dx}\, w(x)$$

that is

$$f(x)\, V(x) - g(x)\, u(x) = \frac{d}{dx}\, w(x)\,.$$

There follows

$$\int_a^{x} f(x)\, V(x)\, dx - \int_a^{x} g(x)\, u(x)\, dx = w(x) - w(a)$$

and therefore, with a passage to the limit for $x \to +\infty$, and using (1.4.22) and (1.4.23):

$$\lim_{x\to+\infty} w(x) = w(a) + \sum_{i=1}^{n}\Bigg[\int_a^{+\infty} f(x)\, v_i(x)\, dx \int_x^{+\infty} g(\xi)\, u_i(\xi)\, d\xi$$

$$- \alpha_i \int_a^{+\infty} g(x)\, u_i(x)\, dx - \int_a^{+\infty} g(x)\, u_i(x)\, dx \int_a^{x} f(\xi)\, v_i(\xi)\, d\xi\Bigg].$$

In virtue of (1.4.19) this formula reduces to

$$\lim_{x\to+\infty} w(x) = w(a) - \sum_{i=1}^{n} \alpha_i \int_a^{+\infty} g(x)\, u_i(x)\, dx$$

and to obtain (1.4.18) we need to prove that the value of this last expression is 0.

To this purpose note that (1.4.12) in connection with (1.4.3) and (1.4.7) gives

$$w(x) = \sum_{h=0}^{n-1} \sum_{i=1}^{n} u_i^{(h)}(x) \left[\alpha_i + \int_a^x f(\xi)\, v_i(\xi)\, d\xi\right] \sum_{j=1}^{n} E^*_{n-h-1}[v_j(x)] \int_x^{+\infty} g(\xi)\, u_j(\xi)\, d\xi$$

$$= \sum_{i=1}^{n} \sum_{j=1}^{n} \left[\alpha_i + \int_a^x f(\xi)\, v_i(\xi)\, d\xi\right] \int_x^{+\infty} g(\xi)\, u_j(\xi)\, d\xi \sum_{h=0}^{n-1} u_i^{(h)}(x)\, E^*_{n-h-1}[v_j(x)]$$

and then using (1.3.24)

$$w(x) = \sum_{i=1}^{n} \sum_{j=1}^{n} \left[\alpha_i + \int_a^x f(\xi)\, v_i(\xi)\, d\xi\right] \int_x^{+\infty} g(\xi)\, u_j(\xi)\, d\xi\, \delta_{ij}$$

$$= \sum_{i=1}^{n} \left[\alpha_i + \int_a^x f(\xi)\, v_i(\xi)\, d\xi\right] \int_x^{+\infty} g(\xi)\, u_i(\xi)\, d\xi\,.$$

For $x = a$ we obtain in particular $w(a) = \sum_{i=1}^{n} \alpha_i \int_a^{+\infty} g(\xi)\, u_i(\xi)\, d\xi$, q.e.d.

In the case of the interval $(-\infty, +\infty)$ the hypotheses analogous to (1.4.5) is

$$g(x)\, u_i(x) \in I\,(-\infty, +\infty)\,, \qquad (i = 1, 2, \ldots, n)\,. \tag{1.4.24}$$

If it is satisfied, we may give in (1.4.4)[1]) to the arbitrary constants β_i the values

$$\beta_i = -\int_{-\infty}^{0} g(\xi)\, u_i(\xi)\, d\xi \quad \text{or} \quad \beta_i = \int_0^{+\infty} g(\xi)\, u_i(\xi)\, d\xi$$

and therefore, with regard to the differential equation (1.4.2) consider the particular solution

$$V_-(x) = -\sum_{i=1}^{n} v_i(x) \int_{-\infty}^{x} g(\xi)\, u_i(\xi)\, d\xi = -\int_{-\infty}^{x} K(\xi, x)\, g(\xi)\, d\xi \tag{1.4.25}$$

[1]) In (1.4.3) and (1.4.4) the symbol a denotes now an arbitrary fixed point; *we shall assume for instance* $a = 0$.

which satisfies the initial conditions null at the point $-\infty$, or the particular solution

$$V_+(x) = \sum_{i=1}^{n} v_i(x) \int_x^{+\infty} g(\xi)\, u_i(\xi)\, d\xi = \int_x^{+\infty} K(\xi, x)\, g(\xi)\, d\xi \qquad (1.4.26)$$

mhich satisfies the initial conditions null at the point $+\infty$.[1])

Then, the theorem analogous to theorem 1.4.I, can be stated as follows:

Theorem 1.4.II. *If the given function* $g(x)$ *of the equation (1.4.2) satisfies the following hypotheses [particular case of (1.4.24)]:*

$$g(x)\, u_i(x) \in L(-\infty, +\infty)\,, \qquad (i = 1, 2, \ldots, n) \qquad (1.4.27)$$

and if the given function $f(x)$ *of the equation (1.4.1) satisfies these other hypotheses*

$$\left.\begin{aligned} &f(x)\, v_i(x) \int_{-\infty}^{x} |g(\xi)\, u_i(\xi)|\, d\xi \in L(-\infty, 0]\,, \\ &f(x)\, v_i(x) \int_{x}^{+\infty} |g(\xi)\, u_i(\xi)|\, d\xi \in L[0, +\infty)\,, \end{aligned}\qquad (i = 1, 2, \ldots, n)\,,\right\} \qquad (1.4.28)$$

then for every solution $u(x)$ *of (1.4.1) and for the particular solutions* $V_-(x)$, $V_+(x)$ *of (1.4.2) above introduced, we have*

$$g(x)\, u(x) \in L(-\infty, +\infty)\,, \qquad (1.4.29)$$

$$f(x)\, V_-(x) \in L(-\infty, 0]\,, \quad f(x)\, V_+(x) \in L[0, +\infty)\,, \qquad (1.4.30)$$

$$\left.\begin{aligned} \lim_{x\to-\infty} w_-(x) &= \lim_{x\to-\infty} \sum_{h=0}^{n-1} u^{(h)}(x)\, E^*_{n-h-1}[V_-(x)] = 0\,, \\ \lim_{x\to+\infty} w_+(x) &= \lim_{x\to+\infty} \sum_{h=0}^{n-1} u^{(h)}(x)\, E^*_{n-h-1}[V_+(x)] = 0\,. \end{aligned}\right\} \qquad (1.4.31)$$

The proof of this theorem is to be carried out reasoning separately on the intervals $(-\infty, 0]$, $[0, +\infty)$ analogously to the proof of theorem 1.4.I. It is only to be noted that the statement (1.4.29) is a consequence of $g(x)\, u(x) \in L(-\infty, 0]$, and $g(x)\, u(x) \in L[0, +\infty)$ which are deduced separately. Note that in the last two theorems, the n hypotheses (1.4.14) or (1.4.27) *generally are not mutually independent*; some may be consequence of the others. The same holds for (1.4.15) or (1.4.28).

[1]) The justification of these two sentences (which are not to be taken literally) is analogous to the one previously given. Instead of (1.4.11) we have now

$$\left.\begin{aligned} &\lim_{x\to-\infty} \sum_{h=0}^{n-1} u_i^{(h)}(x)\, E^*_{n-h-1}[V_-(x)] = 0\,, \\ &\lim_{x\to+\infty} \sum_{h=0}^{n-1} u_i^{(h)}(x)\, E^*_{n-h-1}[V_+(x)] = 0\,, \end{aligned}\right. \qquad (i = 1, 2, \ldots, n).$$

1.5. Problems

1. Let us consider these differential equations:

$$E(u) \equiv \frac{d^n u}{dx^n} = 0 \,, \tag{1}$$

$$E(u) \equiv \sum_{k=0}^{n} a_k \, u^{(n-k)} = 0 \,, \qquad (a_k \text{ constant}) \,, \tag{2}$$

$$E(u) \equiv \sum_{k=0}^{n} \frac{a_k}{(n-k)!} \, \frac{u^{(n-k)}}{x^k} = 0 \,, \qquad (a_k \text{ constant}) \,, \tag{3}$$

and write their adjont equations $E^*(v) = 0$. In each one of the three cases, devise pairs of associated systems $[u_1(x), \ldots, u_n(x)]$, $[v_1(x), \ldots, v_n(x)]$ of solutions of the two equations $E(u) = 0$, $E^*(v) = 0$ (see § 1.3). In the cases (2) and (3) assume that the roots $\alpha_1, \alpha_2, \ldots, \alpha_n$ of the respective characteristic equations

$$\sum_{k=0}^{n} a_k \, \alpha^{n-k} = 0 \,, \qquad \sum_{k=0}^{n} a_k \binom{\alpha}{n-k} = 0$$

be all simple.

2. Prove that, if $U(x) \equiv [u_1(x), \ldots, u_n(x)]$, $V(x) \equiv [v_1(x), \ldots, v_n(x)]$ are two associated systems of solutions of the equations $E(u) = 0$, $E^*(v) = 0$ (see § 1.3), then, fixing arbitrarily a matrix $C = (c_{ik})$, $(i, k = 1, 2, \ldots, n)$, with constant elements and $\det C \neq 0$, there result associated also the systems $\overline{U}(x) = [\overline{u}_1(x), \ldots, \overline{u}_n(x)]$, $\overline{V}(x) = [\overline{v}_1(x), \ldots, \overline{v}_n(x)]$ defined by

$$\overline{U}(x) = V(x)\, C^T \,, \qquad \overline{V}(x) = V(x)\, C^{-1} \,.$$

3. A linear differential operator E of order n:

$$E \equiv \sum_{k=0}^{n} a_k(x) \, \frac{d^{n-k}}{dx^{n-k}} \tag{1}$$

is called *self-adjoint* when there results

$$E^* = (-1)^n \, E \,. \tag{2}$$

In order to construct such an operator we may fix arbitrarily the coefficients $a_{2s}(x) \in A\,C^{n-2s-1}, \left(s = 0, 1, \ldots, \left[\frac{n}{2}\right]\right)$ and then obtain the others coefficients $a_{2r+1}(x)$ by means of

$$a_{2r+1}(x) = \sum_{s=0}^{r} \frac{2^{2r-2s+2}}{r-s+1} B_{2r-2s+2} \binom{n-2s}{2r-2s+1} a_{2s}^{(2r-2s+1)}(x) \,, \tag{3}$$

$$\left(r = 0, 1, \ldots, \left[\frac{n-1}{2}\right]\right),$$

where $B_{2r-2s+2}$ are the Bernouilli's numbers (see § 3.1).

4. Given a function $f(x) \in C^n[a, b]$ and a linear differential operator E of order n [see (1.1.1) and (1.1.2)] we want construct a solution $u(x)$ of the

differential equation $E(u) = 0$ which *interpolate* $f(x)$ in the sense that we must get

$$u^{(k)}(x_i) = f^{(k)}(x_i)\,, \qquad (i = 1, 2, \ldots, m;\ k = 0, 1, \ldots, p_i)\,, \tag{1}$$

with

$$a \leqq x_1 < x_2 < \cdots < x_m \leqq b\,, \quad m \geqq 1\,, \quad p_i \geqq 0\,,$$

$$m + p_1 + p_2 + \cdots + p_m = n\,.$$

Give the condition for the existence and the unicity of $u(x)$ [with any $f(x)$], the expression of $u(x)$ and, using the Green-Lagrange identity (1.2.2), an integral expression of the remainder

$$R(x) = f(x) - u(x)\,. \tag{2}$$

CHAPTER 2

Elementary quadrature formulae and a general procedure for constructing them

2.1. The elementary quadrature formulae on a finite interval.[1]) In this chapter, as in the last, we shall consider exclusively real-valued functions. Let $[a, b]$ be a finite interval on the axis x and n a positive integer; we shall consider an integral of the type

$$\int_a^b g(x)\, u(x)\, dx \tag{2.1.1}$$

where the functions $g(x)$ (*weight functions* which are assumed non zero on a set of positive measure) and $u(x)$ (*argument function*) satisfy the following hypotheses:

$$g(x) \in L[a, b]\,, \qquad u(x) \in A\ C^{n-1}[a, b]\,. \tag{2.1.2}$$

Let us fix in $[a, b]$ a certain number $m \geqq 1$ of points $x_1, x_2, \ldots, x_m$ (*nodes*) in such a way that

$$x_0 = a \leqq x_1 < x_2 < \cdots < x_m \leqq b = x_{m+1}\ ^{2)} \tag{2.1.3}$$

and let us fix also a linear differential operator E of order n [the same n as in (2.1.2)] expressed by a formula of the type (1.1.1) with coefficients which satisfy the hypotheses (1.1.2). We shall call *elementary quadrature formula* of the integral (2.1.1) *relative to the nodes* $x_1, x_2, \ldots, x_m$ *and to the linear differential operator* E, any formula of the type

$$\int_a^b g(x)\, u(x)\, dx = \sum_{h=0}^{n-1} \sum_{i=1}^{m} A_{hi}\, u^{(h)}(x_i) + R[u(x)]\,, \tag{2.1.4}$$

where the constants coefficients A_{hi} *are independent of* $u(x)$ *and the linear functional* $R[u(x)]$ (*remainder*) *is null when* $u(x)$ *is a solution of the homogeneous linear differential equation* $E[u(x)] = 0$.

In other words (2.1.4) is intended to be accompanied by the condition

$$E(u) = 0 \Rightarrow R(u) = 0\,. \tag{2.1.5}$$

In practice, every formula of the type (2.1.4) is used considering the linear functional

$$A(u) = \sum_{h=0}^{n-1} \sum_{i=1}^{m} A_{hi}\, u^{(h)}(x_i) \tag{2.1.6}$$

[1]) See J. Radon [1], A. Ghizzetti [1].

[2]) For sake of uniformity we shall sometimes write x_0 instead of a and x_{m+1} instead of b.

as an approximate value of the integral (2.1.1) and the other functional $R(u)$ as the corresponding error of approximation. The error is null, that is the *quadrature formula is exact*, when $u(x)$ satisfies the differential equation $E(u) = 0$. [1])

In the literature we find innumerable examples of formulae of the type (2.1.4). Most of them use in $A(u)$ only the values $u(x_i)$ of the argument function at the nodes x_i and not (when $n > 1$) values of the derivatives $u'(x_i), \ldots, u^{(n-1)}(x_i)$; but this means only that in such formulae the coefficients $A_{1i}, \ldots, A_{n-1,i}$ turn out to be zero. Moreover almost all classical formulae are exact when $u(x)$ is a polynomial or a trigonometric polynomial; this corresponds to the fact that such formulae are relative to the following particular differential operators:

$E = \frac{d^n}{dx^n}$ (exact formula for polynomials of degree $n - 1$)

$E = \frac{d}{dx} \prod_{k=1}^{\nu} \left(\frac{d^2}{dx^2} + k^2 \right)$ ($n = 2\nu + 1$; exact formula for trigonometric polynomials of order ν).

With this as introduction, we are in a position to arrive quickly at a simple procedure to get a formula of type (2.1.4) under condition (2.1.5).

Let us consider the non-homogeneous linear differential equation

$$E^*[\varphi(x)] = g(x)\,, \tag{2.1.7}$$

where E^* denotes the adjoint operator of E [see (1.1.5)]. Now we fix arbitrarily $m - 1$ solutions

$$\varphi_1(x), \varphi_2(x), \ldots, \varphi_{m-1}(x) \in A\, C^{m-1}[a, b] \tag{2.1.8}$$

of (2.1.7) and consider also the two solutions $\varphi_0(x)$, $\varphi_m(x)$ of the same equation (2.1.7) which are respectively determined by the initial conditions null at the points a, b; namely $\varphi_0^{(h)}(a) = 0$, $\varphi_m^{(h)}(b) = 0$, $(h = 0, 1, \ldots, n - 1)$ and therefore, recalling (1.3.17) and (1.3.18):

$$\varphi_0(x) = -\int_a^x K(\xi, x)\, g(\xi)\, d\xi\,, \qquad \varphi_m(x) = \int_x^b K(\xi, x)\, g(\xi)\, d\xi\,. \tag{2.1.9}$$

We now make use of the Green-Lagrange identity (1.2.2), substituting $v(x) = \varphi_i(x)$, $(i = 0, 1, \ldots, m)$.

Considering that $E^*(\varphi_i) = g$, we get

$$\varphi_i\, E(u) - g\, u = \frac{d}{dx} \sum_{h=0}^{n-1} u^{(h)}\, E^*_{n-h-1}(\varphi_i)\,, \qquad (i = 0, 1, \ldots, m)$$

and therefore, integrating on the interval $[x_i, x_{i+1}]$:

$$\int_{x_i}^{x_{i+1}} g\, u\, dx = -\left[\sum_{h=0}^{n-1} u^{(h)}\, E^*_{n-h-1}(\varphi_i) \right]_{x_i}^{x_{i+1}} + \int_{x_i}^{x_{i+1}} \varphi_i\, E(u)\, dx\,, \qquad (i = 0, 1, \ldots, m)\,.$$

[1]) The solutions of $E(u) = 0$ are not the only functions $u(x)$ for which (2.1.4) is exact; we shall make precise later the whole of such functions (see Remark 4).

Summing with respect to the index i, we arrive at

$$\int_a^b g\,u\,dx = -\sum_{i=0}^{m}\left[\sum_{h=0}^{n-1} u^{(h)}\,E^*_{n-h-1}(\varphi_i)\right]_{x_i}^{x_{i+1}} + \sum_{i=0}^{m}\int_{x_i}^{x_{i+1}} \varphi_i\,E(u)\,dx\,. \quad (2.1.10)$$

Bearing in mind that from $\varphi_0^{(h)}(a) = 0$, $\varphi_m^{(h)}(b) = 0$ follows

$$[E^*_{n-h-1}(\varphi_0)]_{x=x_0} = 0\,, \quad [E^*_{n-h-1}(\varphi_m)]_{x=x_{m+1}} = 0\,, \quad (h = 0, 1, \ldots, n-1)\,,$$

we may then write

$$-\sum_{i=0}^{m}\left[\sum_{h=0}^{n-1} u^{(h)}\,E^*_{n-h-1}(\varphi_i)\right]_{x_i}^{x_{i+1}} = \sum_{h=0}^{n-1}\left\{-\sum_{i=1}^{m+1} u^{(h)}(x_i)\,[E^*_{n-h-1}(\varphi_{i-1})]_{x=x_i}\right.$$

$$\left. + \sum_{i=0}^{m} u^{(h)}(x_i)\,[E^*_{n-h-1}(\varphi_i)]_{x=x_i}\right\} = \sum_{h=0}^{n-1}\sum_{i=1}^{m}\,[E^*_{n-h-1}\,(\varphi_i - \varphi_{i-1})]_{x=x_i}\,u^{(h)}(x_i)\,,$$

so that (2.1.10) can be transformed into

$$\int_a^b g(x)\,u(x)\,dx = \sum_{h=0}^{n-1}\sum_{i=1}^{m}\left\{E^*_{n-h-1}\,[\varphi_i(x) - \varphi_{i-1}(x)]\right\}_{x=x_i} u^{(h)}(x_i)$$

$$+ \sum_{i=0}^{m}\int_{x_i}^{x_{i+1}} \varphi_i(x)\,E[u(x)]\,dx\,. \quad (2.1.11)$$

This formula is precisely of type (2.1.4) with

$$A_{hi} = \{E^*_{n-h-1}\,[\varphi_i(x) - \varphi_{i-1}(x)]\}_{x=x_i}\,, \quad (h = 0, 1, \ldots, n-1;\; i = 1, 2, \ldots, m)\,, \quad (2.1.12)$$

$$R[u(x)] = \sum_{i=0}^{m}\int_{x_i}^{x_{i+1}} \varphi_i(x)\,E[u(x)]\,dx\,, \quad (2.1.13)$$

and from these relations it is evident that the two desired conditions are satisfied [coefficients A_{hi} independent of $u(x)$ and validity of (2.1.5)].

Thus we have the following *rule* for constructing an elementary quadrature formula of type (2.1.4) relative to the nodes $x_1, x_2, \ldots, x_m$ and to the linear differential operator E:

1) *we consider the linear differential equation* $E^*(\varphi) = g$ *and its solutions* $\varphi_0(x)$ *and* $\varphi_m(x)$ *defined by (2.1.9);*

2) *we fix as we like* $m-1$ *other solutions* $\varphi_1(x), \ldots, \varphi_{m-1}(x)$ *of the same equation* $E^*(\varphi) = g$;

3) *we compute the coefficients* A_{hi} *by means of (2.1.12);*

4) *we write the expression for the remainder* $R(u)$ *using (2.1.13).*

This rule is not only reasonably simple; it is also the only one possible, as is proved by the following theorem:

Theorem 2.1.I. *If, having fixed* m *nodes* $x_1, x_2, \ldots, x_m$ *and the* mn *constants* A_{hi}, *the linear functional*

$$R(u) = \int_a^b g\,u\,dx - \sum_{h=0}^{n-1}\sum_{i=1}^{m} A_{hi}\,u^{(h)}(x_i) \quad (2.1.14)$$

is null when u is a solution of the homogeneous linear differential equation $E(u) = 0$, *then there are uniquely determined* $m - 1$ *solutions* $\varphi_1(x), \ldots, \varphi_{m-1}(x)$ *of the differential equation* $E^*(\varphi) = g$ *which, together with the two* $\varphi_0(x)$ *and* $\varphi_m(x)$ *given by* (2.1.9), *make valid* (2.1.12) *and* (2.1.13).

Proof — Equations (2.1.12) may also be written

$$[E^*_{n-h-1}(\varphi_i)]_{x=x_i} = A_{hi} + [E^*_{n-h-1}(\varphi_{i-1})]_{x=x_i}\,,$$
$$(h = 0, 1, \ldots, n-1;\; i = 1, 2, \ldots, m)\,,$$

and therefore, to have them satisfied, it is necessary that $\varphi_1(x)$ be the solution of $E^*(\varphi) = g$ which is characterized by the following initial conditions at the point x_1: [1])

$$[E^*_{n-h-1}(\varphi_1)]_{x=x_1} = A_{h1} + [E^*_{n-h-1}(\varphi_0)]_{x=x_1}\,, \qquad (h = 0, 1, \ldots, n-1)\,,$$

and successively that $\varphi_2(x)$ be the one characterized by the following initial conditions at the point x_2:

$$[E^*_{n-h-1}(\varphi_2)]_{x=x_2} = A_{h2} + [E^*_{n-h-1}(\varphi_1)]_{x=x_2}\,, \qquad (h = 0, 1, \ldots, n-1)\,,$$

and so on for $\varphi_3(x), \ldots, \varphi_{m-1}(x)$. Then we see that $\varphi_m(x)$ must coincide with that solution of $E^*(\varphi) = g$ which satisfies the following initial conditions at the point x_m:

$$[E^*_{n-h-1}(\varphi_m)]_{x=x_m} = A_{hm} + [E^*_{n-h-1}(\varphi_{m-1})]_{x=x_m}\,, \qquad (h = 0, 1, \ldots, m-1)\,.$$

It remains only to see that the solution $\varphi_m(x)$ coincides with that considered in the statement of the theorem, namely that defined by the initial conditions

$$[E^*_{n-h-1}(\varphi_m)]_{x=b} = 0\,, \qquad (h = 0, 1, \ldots, n-1) \qquad (2.1.15)$$

and we have also to prove (2.1.13).

To this purpose we observe that, making use of the solutions $\varphi_1(x), \ldots, \ldots, \varphi_{m-1}(x), \varphi_m(x)$ just introduced, we may do again the same computation which led to (2.1.10). However, we cannot pass from (2.1.10) to (2.1.11) since we do not yet know whether $\varphi_m(x)$ satisfies (2.1.15); therefore instead of (2.1.11) we must write

$$\int_a^b g\, u\, dx = \sum_{h=0}^{n-1} \sum_{i=1}^{m} [E^*_{n-h-1}(\varphi_i - \varphi_{i-1})]_{x=x_i}\, u^{(h)}(x_i)$$
$$- \sum_{h=0}^{n-1} [E^*_{n-h-1}(\varphi_m)]_{x=b}\, u^{(h)}(b) + \sum_{i=0}^{m} \int_{x_i}^{x_{i+1}} \varphi_i\, E(u)\, dx\,. \qquad (2.1.16)$$

Taking into account equation (2.1.14) and the fact that the solutions $\varphi_0, \varphi_1, \ldots \ldots, \varphi_m$ have been chosen so as to satisfy (2.1.12), (2.1.16) may also be written as

$$R(u) = -\sum_{h=0}^{n-1} [E^*_{n-h-1}(\varphi_m)]_{x=b}\, u^{(h)}(b) + \sum_{i=0}^{m} \int_{x_i}^{x_{i+1}} \varphi_i\, E(u)\, dx\,. \qquad (2.1.17)$$

[1]) Please note that, here and in the following, to assign at a point of $[a, b]$ the values of $E^*_{n-h-1}(\varphi)$, $(h = 0, 1, \ldots, n-1)$ is the same thing as to give the values of $\varphi^{(h)}$ and that if the former are all null, the latter too are null and viceversa [see formula (1.3.20)].

We now recall that by hypothesis $R(u)$ must be null when u is a solution of the differential equation $E(u) = 0$. Therefore, denoting by $(u_1, u_2, \ldots, u_n)$ a fundamental system of solutions of that equation, it follows from (2.1.17), putting $u = u_j$, $(j = 1, 2, \ldots, n)$, that

$$\sum_{h=0}^{n-1} u_j^{(h)}(b)\, [E^*_{n-h-1}(\varphi_m)]_{x=b} = 0\,, \qquad (j = 1, 2, \ldots, n)\,.$$

Since the Wronskian det $(u_j^{(h)}(b))$ is different from zero, we conclude that (2.1.15) holds and finally that (2.1.17) reduces to (2.1.13), q.e.d.

On the basis of this theorem, the above stated rule allows us to find as particular cases of (2.1.4) all the known quadrature formulae[1]) and to get several others. We intend to come back to this in § 2.4. Now we add some remarks on the above mentioned rule.

Remark 1. If we wish to be effectively present in the quadrature formula (2.1.4) *all* the prefixed nodes x_i $(i = 1, 2, \ldots, m)$ we require that, for every value of the index i at least one of the coefficients A_{hi} be non-zero. Keeping into account (2.1.12) it is obvious that this is equivalent to requiring that the function $\varphi_i(x) - \varphi_{i-1}(x)$ must not be identically zero. We achieve our objective by *choosing the solution $\varphi_1(x)$ different from $\varphi_0(x)$; $\varphi_2(x)$ different from $\varphi_1(x)$; ...; $\varphi_{m-2}(x)$ different from $\varphi_{m-3}(x)$ and finally $\varphi_{m-1}(x)$ different from $\varphi_{m-2}(x)$ and from $\varphi_m(x)$.*

Remark 2. Having fixed the nodes $x_1, x_2, \ldots, x_m$ and the linear differential operator E, we may write the quadrature formula in $\infty^{(m-1)n}$ different ways, since $(m - 1)\, n$ is the number of arbitrary constants on which the $m - 1$ solutions $\varphi_1(x), \ldots, \varphi_{m-1}(x)$ of the differential equation $E^*(\varphi) = g$ of order n depend.

Remark 3. If $x_1 = a$, (2.1.12) and $[E^*_{n-h-1}(\varphi_0)]_{x=a} = 0$, $(h = 0, 1, \ldots, n - 1)$, imply

$$A_{h1} = [E^*_{n-h-1}\,(\varphi_1 - \varphi_0)]_{x=a} = [E^*_{n-h-1}(\varphi_1)]_{x=a}\,;$$

besides in the expression (2.1.13) for $R(u)$ the first term $\int_a^{x_1} \varphi_0\, E(u)\, dx$ vanishes. Therefore, *if $x_1 = a$, it is useless to consider the solution $\varphi_0(x)$.*

Analogously, if $x_m = b$ we have

$$A_{hm} = [E^*_{n-h-1}\,(\varphi_m - \varphi_{m-1})]_{x=b} = -\, [E^*_{n-h-1}(\varphi_{m-1})]_{x=b}\,,$$

while in $R(u)$ the last term $\int_{x_m}^{b} \varphi_m\, E(u)\, dx$ vanishes; therefore *if $x_m = b$ it is useless to consider the solution $\varphi_m(x)$.*

[1]) At least the most common formulae, which approximate the integral (2.1.1) with a linear functional of the argument function $u(x)$ of type (2.1.6). There exist a few examples of quadrature formulae of other types which do not fall within the theory here stated.

Remark 4. From (2.1.13) it is clear that $R(u) = 0$ for all and only the functions u for which $E(u) \in L[a, b]$ is orthogonal on $[a, b]$ to the so-called *influence function* $\Phi(x)$ defined as

$$\Phi(x) = \varphi_i(x) \quad \text{in} \quad [x_i, x_{i+1}], \quad (i = 0, 1, \ldots, m), \tag{2.1.18}$$

that is, for which

$$E[u(x)] = F(x) - \frac{\int_a^b \Phi(\xi)\, F(\xi)\, d\xi}{\int_a^b \Phi^2(\xi)\, d\xi}\, \Phi(x), \text{[1]} \tag{2.1.19}$$

where $F(x)$ is an *arbitrary* function summable on $[a, b]$. Hence, the *quadrature formula (2.1.4) is exact if and only if the argument function $u(x)$ is the solution of the linear differential equation (2.1.19) where $F(x) \in L[a, b]$ is arbitrary and $\Phi(x)$ is given by (2.1.18)*. Note that the condition (2.1.5) made at the beginning corresponds to the particular choices $F(x) \equiv 0$ or $F(x) = \Phi(x)$.

2.2. The elementary quadrature formulae on the interval $[a, +\infty)$.[2] We shall use here the same terminology and notations as in § 1.4. Let us consider an integral of the type

$$\int_a^{+\infty} g(x)\, u(x)\, dx \tag{2.2.1}$$

where the functions $g(x)$ (*weight function*) and $u(x)$ (*argument function*) satisfy the following hypotheses:

$$g(x) \in L_{\text{loc}}[a, +\infty), \qquad u(x) \in A\, C_{\text{loc}}^{n-1}[a, +\infty), \tag{2.2.2}$$

$$g(x)\, u(x) \in I[a, +\infty). \tag{2.2.3}$$

Let us fix some points $x_1, x_2, \ldots, x_m$ (*nodes*) such that

$$x_0 = a \leqq x_1 < x_2 < \cdots < x_m < +\infty = x_{m+1} \tag{2.2.4}$$

and a linear differential operator E of order n expressed by a formula of the type (1.1.1) where the coefficients $a_k(x)$ satisfy the hypotheses

$$a_0(x) = 1; \quad a_k(x) \in A\, C_{\text{loc}}^{n-k-1}[a, +\infty), \quad (k = 1, 2, \ldots, n-1);$$
$$a_n(x) \in L_{\text{loc}}[a, +\infty). \tag{2.2.5}$$

We shall call *elementary quadrature formula* for the *convergent integral (2.2.1) relative to nodes $x_1, x_2, \ldots, x_m$ and to the linear differential operator E*, any formula of the type

$$\int_a^{+\infty} g(x)\, u(x)\, dx = \sum_{h=0}^{n-1} \sum_{i=1}^{m} A_{hi}\, u^{(h)}(x_i) + R[u(x)] \tag{2.2.6}$$

[1]) We have $\int_a^b \Phi^2(\xi)\, d\xi > 0$ since the case in which every solution $\varphi_i(x)$ of the equation $E^*(\varphi) = g$ may be identically zero in $[x_i, x_{i+1}]$ is excluded since the weight function $g(x)$ was supposed different from zero on a set of positive measure.

[2]) See A. Ghizzetti [2]; [3], [4].

whose coefficients A_{hi} are independent of $u(x)$ and such that

$$E(u) = 0 \Rightarrow R(u) = 0 \,. \tag{2.2.7}$$

To construct such a formula it is sufficient to give a rule analogous to the one of § 2.1, however with some clarifications and remarks.

If we introduce a fundamental system $(u_1, u_2, \ldots, u_n)$ of the solutions of $E(u) = 0$, it is obvious that in order that the condition (2.2.7) make sense, the hypotheses

$$g(x)\, u_i(x) \in I[a, +\infty)\,, \qquad (i = 1, 2, \ldots, n)\,, \tag{2.2.8}$$

is indispensable.

With this condition established, and fixing one point b with $b > x_m$, we may perform, with respect to the interval $[a, b]$, a computation analogous to that of § 2.1.

Namely, we denote by $\varphi_0(x)$ that solution of the differential equation $E^*[\varphi(x)] = g(x)$ which satisfies the initial conditions null at the point $x = a$ and by $\varphi_1(x), \ldots, \varphi_m(x)$ m other solutions arbitrarily fixed.

We apply the Green-Lagrange identity (1.2.2), putting $v(x) = \varphi_i(x)$ $(i = 0, 1, \ldots, m)$. This leads to

$$\varphi_i\, E(u) - g\, u = \frac{d}{dx} \sum_{h=0}^{n-1} u^{(h)}\, E^*_{n-h-1}(\varphi_i)\,, \qquad (i = 0, 1, \ldots, m-1, m)\,,$$

from which, integrating them on the intervals $[x_i, x_{i+1}]$ $(i = 0, 1, \ldots, m-1)$, $[x_m, b]$ respectively and summing the resulting equations, we obtain

$$\int_a^b g\, u\, dx = -\sum_{i=0}^{m-1} \left[\sum_{h=0}^{n-1} u^{(h)}\, E^*_{n-h-1}(\varphi_i)\right]_{x_i}^{x_{i+1}} - \left[\sum_{h=0}^{n-1} u^{(h)}\, E^*_{n-h-1}(\varphi_m)\right]_{x_m}^{b}$$
$$+ \sum_{i=0}^{m-1} \int_{x_i}^{x_{i+1}} \varphi_i\, E(u)\, dx + \int_{x_m}^{b} \varphi_m\, E(u)\, dx\,,$$

from which, with a simple transformation and use of the fact that $\varphi_0^{(h)}(a) = 0$, $(h = 0, 1, \ldots, n-1)$,

$$\int_a^b g\, u\, dx = \sum_{h=0}^{n-1} \sum_{i=1}^{m} [E^*_{n-h-1}(\varphi_i - \varphi_{i-1})]_{x=x_i}\, u^{(h)}(x_i)$$
$$- \left[\sum_{h=0}^{n-1} u^{(h)}\, E^*_{n-h-1}(\varphi_m)\right]_{x=b} + \sum_{i=0}^{m-1} \int_{x_i}^{x_{i+1}} \varphi_i\, E(u)\, dx + \int_{x_m}^{b} \varphi_m\, E(u)\, dx\,. \tag{2.2.9}$$

This relation, by means of a passage to the limit with $b \to +\infty$, will evidently lead to a formula of the type (2.2.6), *if it is possible to choose* $\varphi_m(x)$ *in such a way that*

$$\lim_{b \to +\infty} \left[\sum_{h=0}^{n-1} u^{(h)}\, E^*_{n-h-1}(\varphi_m)\right]_{x=b} = 0\,, \qquad \varphi_m\, E(u) \in I[a, +\infty)\,. \tag{2.2.10}$$

It is possible to construct examples (see A. Ghizzetti [2]) which show that (2.2.3) and (2.2.8) are generally not sufficient to assure that the differential equation $E^*(\varphi) = g$ possesses a solution satisfying (2.2.10). On the other hand it does not seem easy, confining ourselves to the field of integrals which are simply convergent and avoiding any particularization of the operator E, to arrive at a formulation of hypotheses which can assure the validity of the procedure but not be too restrictive. Therefore we think it appropriate to confine ourselves to *giving a general theorem for the case of absolutely convergent integrals.*

In the case of simply convergent integrals, we think it better to study (2.2.10) case by case, that is *for any fixed choice of the operator E.*

The theorem for absolutely convergent integrals is immediately derived from theorem 1.4.I and can be stated as follows:

Theorem 2.2.I. *Under hypotheses (2.2.2) and letting* $(u_1, \ldots, u_n)$, $(v_1, \ldots, v_n)$ *be two associated systems of solutions of the differential equations* $E(u) = 0$, $E^*(v) = 0$, *we assume moreover that the weight function* $g(x)$ *satisfy the following hypotheses*

$$g(x)\, u_i(x) \in L[a, +\infty)\,, \qquad (i = 1, 2, \ldots, n)\,, \tag{2.2.11}$$

and that the argument function $u(x)$ *satisfy also*

$$E[u(x)]\, v_i(x) \int_x^{+\infty} |g(\xi)\, u_i(\xi)|\, d\xi \in L[a, +\infty)\,, \qquad (i = 1, 2, \ldots, n)\,. \tag{2.2.12}$$

Then

$$g(x)\, u(x) \in L[a, +\infty) \tag{2.2.13}$$

and, if as solution $\varphi_m(x)$ *of* $E^*(\varphi) = g$ *we take that* $V(x)$ *which satisfies the initial conditions null at the point* $+\infty$ *[see (1.4.9)], (2.2.10) is satisfied and (2.2.6) follows from (2.2.9) together with*

$$A_{hi} = [E^*_{n-h-1}(\varphi_i - \varphi_{i-1})]_{x=x_i}\,, \quad (h = 0, 1, \ldots, n-1;\; i = 1, 2, \ldots, m)\,, \tag{2.2.14}$$

$$R(u) = \sum_{i=0}^{m} \int_{x_i}^{x_{i+1}} \varphi_i\, E(u)\, dx\,, \qquad (\text{with } x_0 = a,\; x_{m+1} = +\infty)\,. \tag{2.2.15}$$

We have then the following *rule* to construct an elementary quadrature formula of type (2.2.6) relative to the nodes $x_1, x_2, \ldots, x_m$ and to the linear differential operator E [when for instance the hypotheses (2.2.2), (2.2.11) and (2.2.12) are satisfied]:

1) *we consider the linear differential equation* $E^*(\varphi) = g$ *and its solutions*

$$\varphi_0(x) = -\int_a^x K(\xi, x)\, g(\xi)\, d\xi\,, \qquad \varphi_m(x) = \int_x^{+\infty} K(\xi, x)\, g(\xi)\, d\xi\,; \tag{2.2.16}$$

2) *we fix arbitrarily* $m-1$ *other solutions* $\varphi_1(x), \ldots, \varphi_{m-1}(x)$ *of the same equation* $E^*(\varphi) = g$;

3) *we compute the coefficients* A_{hi} *by means of (2.2.14);*

4) *we write the expression of the remainder* $R(u)$ *using (2.2.15).*

As in the preceding § 2.1 this rule is essentially the only possible one, as shown by the following theorem:

Theorem 2.2.II. *Having fixed the linear differential operator E, we assume that the functions $g(x)$, $u(x)$ satisfy the hypotheses (2.2.2), (2.2.11) and (2.2.12). Having then fixed the m nodes $x_1, x_2, \ldots, x_m$ and the mn constants A_{hi}, if the linear functional*

$$R(u) = \int_a^{+\infty} g\, u\, dx - \sum_{h=0}^{n-1} \sum_{i=1}^{m} A_{hi}\, u^{(h)}(x_i) \tag{2.2.17}$$

is null when u is a solution of the differential equation $E(u) = 0$, then $m - 1$ solutions $\varphi_1(x), \ldots, \varphi_{m-1}(x)$ of the differential equation $E^*(\varphi) = g$ are uniquely determined, which, together with the two solutions $\varphi_0(x)$ and $\varphi_m(x)$ defined by (2.2.16), imply (2.2.14) and (2.2.15).

Proof. As in the proof of theorem 2.1.I, first we see that in order that (2.2.14) hold, the above mentioned solutions $\varphi_1(x), \varphi_2(x), \ldots, \varphi_{m-1}(x)$ must be necessarily those determined by the following initial conditions respectively at the points $x_1, x_2, \ldots, x_{m-1}$;

$$\left.\begin{aligned} [E^*_{n-h-1}(\varphi_1)]_{x=x_1} &= A_{h1} + [E^*_{n-h-1}(\varphi_0)]_{x=x_1}\,, \\ [E^*_{n-h-1}(\varphi_2)]_{x=x_2} &= A_{h2} + [E^*_{n-h-1}(\varphi_1)]_{x=x_2}\,, \\ &\cdots\cdots\cdots\cdots\cdots \\ [E^*_{n-h-1}(\varphi_{m-1})]_{x=x_{m-1}} &= A_{h,m-1} + [E^*_{n-h-1}(\varphi_{m-2})]_{x=x_{m-1}}\,, \end{aligned}\right\} \tag{2.2.18}$$
$$(h = 0, 1, \ldots, n-1)\,,$$

and moreover that the solution φ_m must coincide with the one determined by these initial conditions at the point x_m:

$$[E^*_{n-h-1}(\varphi_m)]_{x=x_m} = A_{hm} + [E^*_{n-h-1}(\varphi_{m-1})]_{x=x_m}\,, \qquad (h = 0, 1, \ldots, n-1)\,. \tag{2.2.19}$$

With the solutions $\varphi_1, \ldots, \varphi_m$ of $E^*(\varphi) = g$ defined by (2.2.18) and (2.2.19) we may write (2.2.9) and pass to the limit as $b \to +\infty$. In virtue of theorem 1.4.I, hypotheses (2.2.11), (2.2.12) assure that $g(x)\, u(x) \in L[a, +\infty)$; the passage to the limit in (2.2.9) yields, considering also that we have already proved the validity of (2.2.14):

$$\int_a^{+\infty} g\, u\, dx = \sum_{h=0}^{n-1} \sum_{i=1}^{m} A_{hi}\, u^{(h)}(x_i) + \sum_{i=0}^{m-1} \int_{x_i}^{x_{i+1}} \varphi_i\, E(u)\, dx$$
$$+ \lim_{b \to +\infty} \left\{ - \left[\sum_{h=0}^{n-1} u^{(h)}\, E^*_{n-h-1}(\varphi_m) \right]_{x=b} + \int_{x_m}^{b} \varphi_m\, E(u)\, dx \right\},$$

that is for (2.2.17)

$$R(u) = \sum_{i=0}^{m-1} \int_{x_i}^{x_{i+1}} \varphi_i\, E(u)\, dx + \lim_{b \to +\infty} \left\{ - \left[\sum_{h=0}^{n-1} u^{(h)}\, E^*_{n-h-1}(\varphi_m) \right]_{x=b} + \int_{x_m}^{b} \varphi_m\, E(u)\, dx \right\}. \tag{2.2.20}$$

Since $R(u)$ must be zero when u coincides with the solutions $u_1, \ldots, u_n$ of $E(u) = 0$, we find

$$\lim_{x\to+\infty} \sum_{h=0}^{n-1} u_j^{(h)}(x)\, E^*_{n-h-1}[\varphi_m(x)] = 0\,, \qquad (j = 1, 2, \ldots, n)\,. \tag{2.2.21}$$

Taking for $\varphi_m(x)$ the representation [see (1.4.4)]

$$\varphi_m(x) = \sum_{i=1}^{n} v_i(x) \left[\beta_i - \int_a^x g(\xi)\, u_i(\xi)\, d\xi\right], \tag{2.2.22}$$

the value to the limit given in (2.2.21) is

$$\lim_{x\to+\infty} \sum_{h=0}^{n-1} u_j^{(h)}(x) \sum_{i=1}^{n} E^*_{n-h-1}[v_i(x)] \left[\beta_i - \int_a^x g(\xi)\, u_i(\xi)\, d\xi\right]$$

$$= \lim_{x\to+\infty} \sum_{i=1}^{n} \left[\beta_i - \int_a^x g(\xi)\, u_i(\xi)\, d\xi\right] \sum_{h=0}^{n-1} u_j^{(h)}(x)\, E^*_{n-h-1}[v_i(x)]\,.$$

That is, on the basis of theorem 1.3.III,

$$\lim_{x\to+\infty} \sum_{i=1}^{n} \left[\beta_i - \int_a^x g(\xi)\, u_i(\xi)\, d\xi\right] \delta_{ij} = \beta_j - \int_a^{+\infty} g(\xi)\, u_j(\xi)\, d\xi\,,$$

therefore we must have, for (2.2.21)

$$\beta_j = \int_a^{+\infty} g(\xi)\, u_j(\xi)\, d\xi\,, \qquad (j = 1, 2, \ldots, n)\,.$$

Substituting this expression for β_i in (2.2.22) we deduce that for the solution $\varphi_m(x)$ we have

$$\varphi_m(x) = \sum_{i=1}^{n} v_i(x) \int_x^{+\infty} g(\xi)\, u_i(\xi)\, d\xi\,.$$

Now we see from (1.4.8) and (1.4.9) that $\varphi_m(x)$ is exactly that solution which satisfies the initial conditions null at the point $+\infty$. With the notations of theorem 1.4.I we have namely $\varphi_m(x) = V(x)$; using then the two other conclusions of that theorem, we conclude that

$$\lim_{b\to+\infty} \left[\sum_{h=0}^{n-1} u^{(h)}\, E^*_{n-h-1}(\varphi_m)\right]_{x=b} = 0\,,$$

$$\lim_{b\to+\infty} \int_{x_m}^{b} \varphi_m\, E(u)\, dx = \int_{x_m}^{+\infty} \varphi_m\, E(u)\, dx\,,$$

therefore (2.2.20) reduces to (2.2.15) q.e.d.

Note that formulae (2.2.6), (2.2.14) and (2.2.15) in the case of the interval $[a, +\infty)$ are quite similar to (2.1.4), (2.1.12) and (2.1.13) relative to the case of the finite interval $[a, b]$. Remarks 1, 2, and 3 of the preceding § 2.1 hold here too. Remark 3 is to be limited to the fact that if $x_1 = a$ it is useless to consider the solution $\varphi_0(x)$. As to Remark 4, it still stands that (2.2.6) is exact for all and only the functions $u(x)$ for which $E(u)$ is in $[a, +\infty)$ orthogonal to the influence function $\Phi(x)$ determined by (2.1.18), but generally (2.1.19) (with $b = +\infty$) has no meaning.

2.3. Elementary quadrature formulae on the interval $(-\infty, +\infty)$.[1]) Let us consider an integral of the type

$$\int_{-\infty}^{+\infty} g(x)\, u(x)\, dx \tag{2.3.1}$$

where the functions $g(x)$ and $u(x)$ satisfy the hypotheses

$$g(x) \in L_{\text{loc}}(-\infty, +\infty)\,, \qquad u(x) \in A\, C_{\text{loc}}^{n-1}(-\infty, +\infty)\,, \tag{2.3.2}$$

$$g(x)\, u(x) \in I(-\infty, +\infty)\,. \tag{2.3.3}$$

Having fixed the *nodes* $x_1, x_2, \ldots, x_m$ such as to have the ordering

$$x_0 = -\infty < x_1 < x_2 < \cdots < x_m < +\infty = x_{m+1} \tag{2.3.4}$$

and a *linear differential operator* E of order n, of type (1.1.1) and whose coefficients $a_k(x)$ satisfy the hypotheses

$$a_0(x) = 1; \quad a_k(x) \in A\, C_{\text{loc}}^{n-k-1}(-\infty, +\infty), \qquad (k = 1, 2, \ldots, n-1);$$
$$a_n(x) \in L_{\text{loc}}(-\infty, +\infty) \tag{2.3.5}$$

we may find a corresponding *elementary quadrature formula*

$$\int_{-\infty}^{+\infty} g(x)\, u(x)\, dx = \sum_{h=0}^{n-1} \sum_{i=1}^{m} A_{hi}\, u^{(h)}(x_i) + R[u(x)] \tag{2.3.6}$$

under the usual condition

$$E(u) = 0 \Rightarrow R(u) = 0\,. \tag{2.3.7}$$

Keeping the notation of § 1.4, we may reason as in § 2.2 requiring [together with (2.3.2)] the hypotheses

$$\left.\begin{aligned} & g(x)\, u_i(x) \in L(-\infty, +\infty)\,, \\ & E[u(x)]\, v_i(x) \int_{-\infty}^{x} |g(\xi)\, u_i(\xi)|\, d\xi \in L(-\infty, 0]\,, \\ & E[u(x)]\, v_i(x) \int_{x}^{+\infty} |g(\xi)\, u_i(\xi)|\, d\xi \in L[0, +\infty)\,, \\ & (i = 1, 2, \ldots, n)\,, \end{aligned}\right\} \tag{2.3.8}$$

which imply (by theorem 1.4.II) $g(x)\, u(x) \in L(-\infty, +\infty)$. More precisely, having fixed $a < x_1$ and $b > x_m$ and choosing arbitrarily $m + 1$ solutions $\varphi_0(x), \varphi_1(x), \ldots, \varphi_m(x)$ of $E^*(\varphi) = g$, we may write the following formula, similar to (2.2.9):

$$\int_a^b g\, u\, dx = \left[\sum_{h=0}^{n-1} u^{(h)}\, E_{n-h-1}^*(\varphi_0)\right]_{x=a} + \sum_{h=0}^{n-1} \sum_{i=1}^{m} [E_{n-h-1}^*(\varphi_i - \varphi_{i-1})]_{x=x_i}\, u^{(h)}(x_i) -$$
$$- \left[\sum_{h=0}^{n-1} u^{(h)}\, E_{n-h-1}^*(\varphi_m)\right]_{x=b} + \int_a^{x_1} \varphi_0\, E(u)\, dx + \sum_{i=1}^{m-1} \int_{x_i}^{x_{i+1}} \varphi_i\, E(u)\, dx + \int_{x_m}^{b} \varphi_m\, E(u)\, dx \tag{2.3.9}$$

[1]) See A. Ghizzetti [2], [3], [4].

and take the limits as $a \to -\infty$, $b \to +\infty$. Applying theorem 1.4.II we see that with such passages to the limit, (2.3.9) yields a formula of the type (2.3.6) if we take as $\varphi_0(x)$, $\varphi_m(x)$ the solutions of $E^*(\varphi) = g$ which satisfy the initial conditions null respectively at the points $-\infty$, $+\infty$.

We thus arrive to the following rule, similar to the one of § 2.2, and valid when the hypotheses (2.3.2) and (2.3.8) are satisfied:

1) *we consider the linear differential equation* $E^*(\varphi) = g$ *and its solution* $\varphi_0(x)$ *defined by the initial conditions null at the point* $-\infty$ *and thus given by*

$$\varphi_0(x) = -\int_{-\infty}^{x} K(\xi, x)\, g(\xi)\, d\xi \tag{2.3.10}$$

and the solution $\varphi_m(x)$ *which satisfies the initial conditions null at the point* $+\infty$, *expressed by*

$$\varphi_m(x) = \int_{x}^{+\infty} K(\xi, x)\, g(\xi)\, d\xi\,; \tag{2.3.11}$$

2) *we fix as we like* $m-1$ *solutions* $\varphi_1(x), \ldots, \varphi_{m-1}(x)$ *of the same equation* $E^*(\varphi) = g$;

3) *we compute the coefficients* A_{hi} *by means of*

$$A_{hi} = [E^*_{n-h-1}(\varphi_i - \varphi_{i-1})]_{x=x_i}, \qquad (h = 0, 1, \ldots, n-1;\ i = 1, 2, \ldots, m)\,; \tag{2.3.12}$$

4) *we write the expression of the remainder* $R(u)$ *in the following way*:

$$R(u) = \sum_{i=0}^{m} \int_{x_i}^{x_{i+1}} \varphi_i\, E(u)\, dx\,, \quad (\text{with } x_0 = -\infty,\ x_{m+1} = +\infty)\,. \tag{2.3.13}$$

This rule is essentially the only possible since there exists a theorem similar to theorem 2.2.II whose statement and proof we leave to the reader.

Finally Remarks 1, 2, 4 of § 2.1 hold, if we remember that (2.1.19) (with $a = -\infty$, $b = +\infty$) has no longer any meaning.

2.4. Some remarks on the rule for obtaining quadrature formulae. If we examine the three rules given in § 2.1, 2.2, and 2.3 for obtaining elementary quadrature formulae on a finite interval $[a, b]$, on an interval $[a, +\infty)$ and on the interval $(-\infty, +\infty)$, we see that they are formally identical, so that we may always refer to (2.1.4), (2.1.9), (2.1.12), (2.1.13) with the understanding that a can also be finite and $b = +\infty$, or $a = -\infty$, $b = +\infty$. The only difference among the three cases is in the formulation of the necessary hypotheses regarding the weight function $g(x)$ and the argument function $u(x)$. In the case of infinite intervals these hypotheses call in also two associated systems $[u_1(x), \ldots, u_n(x)]$, $[v_1(x), \ldots, v_n(x)]$ of solutions of the differential equations $E(u) = 0$, $E^*(v) = 0$. We know that the rules given are the most general possible, wherefrom we may derive systematically *all the known quadrature formulae, choosing appropriately* case by case the nodes $x_1, \ldots, x_m$ and the linear differential operator E.

Such deductions substantially do not present anything new and generally are of no difficulty.[1])

On the contrary we may say that in the set of all quadrature formulae supplied by the above given systematic procedure, the classical ones are of no particular importance. We may also suggest that it is better to make the most of the procedure by getting, case by case, a quadrature formula particularly suited to the calculation of a given integral $\int_a^b F(x)\,dx$ with $F(x) \in L[a, b]$ taking advantage, not only from the arbitrary choice of the nodes $x_1, \ldots, x_m$ and of the linear differential operator E, of order n, but *also from the numberless ways we may write* $F(x) = g(x)\,u(x)$ *with* $g(x) \in L[a, b]$ *and* $u(x) \in A\,C^{n-1}[a, b]$.

This point of view is not yet widespread and therefore, for the convenience of the reader, we shall gather in Chapter 4 several examples of deductions of classical formulae, together with others that we consider new. For this purpose it will be necessary to state in Chapter 3 some notes on certain auxiliary elements (Bernoulli numbers and polynomials, orthogonal polynomials, etc.). In applying the general procedure it is useful to keep in mind some observations which can be of great utility, above all when we try to obtain quadrature formulae with certain pre-established qualifications (for instance that certain coefficients A_{hi} be zero). We shall present these observations in the remainder of this § 2.4 and in the two successive ones.

The given rule for obtaining the formula

$$\int_a^b g(x)\,u(x)\,dx = \sum_{h=0}^{n-1} \sum_{i=1}^{m} A_{hi}\, u^{(h)}(x_i) + R(u)\,, \qquad [E(u) = 0 \Rightarrow R(u) = 0]\,, \tag{2.4.1}$$

with a, b finite or infinite, considers first the solutions $\varphi_0(x), \varphi_1(x), \ldots, \varphi_m(x)$ of the differential equation $E^*(\varphi) = g$, then the computation of the coefficients A by means of

$$A_{hi} = [E^*_{n-h-1}(\varphi_i - \varphi_{i-1})]_{x=x_i} \tag{2.4.2}$$

and finally the expression of the remainder written

$$R(u) = \sum_{i=0}^{m} \int_{x_i}^{x_{i+1}} \varphi_i\, E(u)\,dx\,. \tag{2.4.3}$$

Sometimes it is more convenient to compute first the coefficients A_{hi}, considering n linearly independent solutions $u_1(x), \ldots, u_n(x)$ of the linear differential equation $E(u) = 0$ and writing the system of n linear equations

[1]) Some difficulties may arise when we try to deduce from the expression of the remainder $R(u)$ formulae of majoration and evaluation of the remainder itself. In such operations it is often necessary to study the sign of the influence function $\Phi(x)$ then making use of very delicate technique.

in mn unknowns A_{hi};

$$\int_a^b g(x)\, u_j(x)\, dx = \sum_{h=0}^{n-1} \sum_{i=1}^{m} A_{hi}\, u_j^{(h)}(x)\,, \qquad (j = 1, 2, \ldots, n)\,. \tag{2.4.4}$$

From this system we may derive for instance the n coefficients $A_{01}, A_{11}, \ldots, A_{n-1,1}$ in terms of the remaining $(m-1)\,n$ which can be fixed arbitrarily (see § 2.1, Remark 2) or with any particular criteria.
Furthermore we may apply the following theorem:

Theorem 2.4.I. *Assuming that the coefficients A_{hi} of (2.4.1) are known, the functions $\varphi_0(x), \varphi_1(x), \ldots, \varphi_m(x)$ necessary to write (2.4.3), are given by the formula*

$$\varphi_i(x) = -\int_a^x K(\xi, x)\, g(\xi)\, d\xi + \sum_{h=0}^{n-1} \sum_{j=1}^{i} A_{hj} \left[\frac{\partial^h}{\partial \xi^h} K(\xi, x)\right]_{\xi = x_j}, \tag{2.4.5}$$

$$(i = 0, 1, \ldots, m)$$

where $K(x, \xi)$ is the resolvent kernel relative to the operator E [see (1.3.4)].

Proof. For $i = 0$ equation (2.4.5) is true since it coincides with the first equation of (2.1.9); for $i = m$ it cannot immediately seen that (2.4.5) coincides with the second one of (2.1.9), that is that

$$-\int_a^x K(\xi, x)\, g(\xi)\, d\xi + \sum_{h=0}^{n-1} \sum_{j=1}^{m} A_{hj} \left[\frac{\partial^h}{\partial \xi^h} K(\xi, x)\right]_{\xi = x_j} = \int_x^b K(\xi, x)\, g(\xi)\, d\xi\,. \tag{2.4.6}$$

But it can easily proved because if in (2.4.1) we denote by ξ the variable of integration and put $u(\xi) = K(\xi, x)$ [solution of $E(u) = 0$, whence $R(u) = 0$], we obtain

$$\int_a^b g(\xi)\, K(\xi, x)\, d\xi = \sum_{h=0}^{n-1} \sum_{i=1}^{m} A_{hi} \left[\frac{\partial^h}{\partial \xi^h} K(\xi, x)\right]_{\xi = x_i}$$

which is equivalent to (2.4.6).

In order to finish the proof of (2.4.5) it will be sufficient to prove that (2.4.2) is satisfied. For this purpose we observe that from (2.4.5) we may derive

$$\varphi_i(x) - \varphi_{i-1}(x) = \sum_{h=0}^{n-1} A_{hi} \left[\frac{\partial^h}{\partial \xi^h} K(\xi, x)\right]_{\xi = x}$$

and therefore for (1.3.4)

$$\varphi_i(x) - \varphi_{i-1}(x) = \sum_{h=0}^{n-1} A_{hi} \sum_{j=1}^{n} u_j^{(h)}(x_i)\, v_j(x);$$

it follows, using (1.3.9), that

$$[E^*_{n-k-1}(\varphi_i - \varphi_{i-1})]_{x=x_i} = \sum_{h=0}^{n-1} A_{hi} \sum_{j=1}^{n} u_j^{(h)}(x_i)\, [E^*_{n-k-1}(v_j)]_{x=x_i}$$
$$= \sum_{h=0}^{n-1} A_{hi}\, \delta_{hk} = A_{ki}\,,$$

which is (2.4.2) q.e.d.

We next note that in (2.4.1) the condition

$$E(u) = 0 \Rightarrow R(u) = 0 \tag{2.4.7}$$

can obviously be replaced by

$$E(u) = E(\bar{u}) \Rightarrow R(u) = R(\bar{u})\,, \tag{2.4.8}$$

where $\bar{u}(x)$ denotes another function satisfying the same hypotheses of $u(x)$.

Finally note that (2.4.7) requires that the quadrature formula (2.4.1) be exact for the functions $u(x)$ belonging to a certain n-dimensional linear space. Instead of defining the space by means of the operator E, we can define it by its linearly independent n points; *namely (2.4.7) can be replaced by the condition that the quadrature formula be exact for n assigned functions $u_1(x), \ldots, u_n(x)$ having Wronskian $U(x) \neq 0$ in the interval of integration.* We then have the result

$$E(u) = \frac{(-1)^n}{U(x)} \begin{vmatrix} u & u' & \ldots & u^{(n-1)} & u^{(n)} \\ u_1 & u_1' & \ldots & u_1^{(n-1)} & u_1^{(n)} \\ u_2 & u_2' & \ldots & u_2^{(n-1)} & u_2^{(n)} \\ \cdot & \cdot & \cdot & \cdot & \cdot \\ u_n & u_n' & \ldots & u_n^{(n-1)} & u_n^{(n)} \end{vmatrix}\,. \tag{2.4.9}$$

2.5. Gauss problem.[1]) Let us consider again the elementary quadrature formula

$$\int_a^b g(x)\, u(x)\, dx = \sum_{h=0}^{n-1} \sum_{i=1}^{m} A_{hi}\, u^{(h)}(x_i) + R(u)\,, \quad [E(u) = 0 \Rightarrow R(u) = 0]\,. \tag{2.5.1}$$

In its right member we find the values, at the nodes $x_1, x_2, \ldots, x_m$, of the function u and of its derivatives up to order $n - 1$; we know (§ 2.1, Remark 2) that in the coefficients A_{hi} there are altogether $(m - 1)\, n$ arbitrary parameters. Therefore the question is whether, having fixed an integer p (with $1 \leqq p \leqq n - 1$), it is possible to make use of the arbitrary nature of these parameters to drop from the formula the values $u^{(h)}(x_i)$ of the derivatives of order higher than $n - p - 1$, that is whether there can exist a formula of the type

$$\int_a^b g(x)\, u(x)\, dx = \sum_{h=0}^{n-p-1} \sum_{i=1}^{m} A_{hi}\, u^{(h)}(x_i) + R(u)\,, \quad [E(u) = 0 \Rightarrow R(u) = 0]\,. \tag{2.5.2}$$

[1]) See A. Ghizzetti [1].

If $(u_1, u_2, \ldots, u_n)$ are a fundamental system of solutions of $E(u) = 0$ it is clear that a necessary and sufficient condition that (2.5.2) hold is that the $m(n-p)$ coefficients A_{hi} satisfy the following system of n linear equations

$$\sum_{h=0}^{n-p-1} \sum_{i=1}^{m} A_{hi}\, u_j^{(h)}(x_i) = \int_a^b g(x)\, u_j(x)\, dx\,, \qquad (j = 1, 2, \ldots, n)\,. \tag{2.5.3}$$

To discuss this system of equations, let us consider the transposed homogeneous system which, denoting by $c_1, c_2, \ldots, c_n$ the relevant unknowns, may be written

$$\sum_{j=1}^{n} u_j^{(h)}(x_i)\, c_j = 0\,, \qquad (h = 0, 1, \ldots, n-p-1; \quad i = 1, 2, \ldots, m)\,. \tag{2.5.4}$$

If the rank of the matrix $(u_j^{(h)}(x_i))$ with $m(n-p)$ rows and n columns, is equal to n, [1] the system (2.5.4) has no non-trivial solutions and therefore the system (2.5.3) is certainly consistent and has $\infty^{m(n-p)-n}$ solutions.

Let us now assume that the matrix has rank $n - q$ (with $q \geqq 1$). [2] In this case the system (2.5.4) has q linearly independent non-trivial solutions $(c_{1r}, c_{2r}, \ldots, c_{nr})$, $(r = 1, 2, \ldots, q)$ and by a well-known criterion the system (2.5.3) is consistent if and only if the q conditions

$$\sum_{j=1}^{n} c_{jr} \int_a^b g(x)\, u_j(x)\, dx = 0\,, \qquad (r = 1, 2, \ldots, q); \tag{2.5.5}$$

are satisfied.

In this case (2.5.3) has $\infty^{m(n-p)-(n-q)}$ solutions. Now we shall try to present these results in a more illuminating form. To this purpose we shall consider the following problem: determine a solution of the differential equation $E(u) = 0$ which together with its derivatives up to the order $n - p - 1$ is zero at each of the points $x_1, x_2, \ldots, x_m$.

This is an *homogeneous boundary problem* expressed by the formula

$$E(u) = 0; \quad u^{(h)}(x_i) = 0\,, \qquad (h = 0, 1, \ldots, n-p-1; \quad i = 1, 2, \ldots, m)\,. \tag{2.5.6}$$

Since the general solution of $E(u) = 0$ is given by $u(x) = \sum_{j=1}^{n} c_j\, u_j(x)$, with c_j arbitrary constants, the boundary conditions are expressed by the following system of $m(n-p)$ equations in n unknowns $c_1, \ldots, c_n$:

$$\sum_{j=1}^{n} c_j\, u_j^{(h)}(x_i) = 0\,, \qquad (h = 0, 1, \ldots, n-p-1; \; i = 1, 2, \ldots, m)\,,$$

which coincides with the system (2.5.4).

[1]) This is possible only if $n \leqq m(n-p)$.

[2]) We must have $n - q \leqq m(n-p)$. Moreover the matrix considered contains the m matrices $(n-p) \times n$ formed by the first $n - p$ rows of the Wronskian of $u_1, \ldots, u_n$ computed at the points x_i $(i = 1, 2, \ldots, m)$; these matrices have rank $n - p$ and therefore we certainly have $n - q \geqq n - p$. In summary, we may write

$$n - p \leqq n - q \leqq m(n-p)\,.$$

Therefore, saying that the system (2.5.4) has no non-trivial solutions is equivalent to saying that the problem (2.5.6) admits only the solution $u(x) = 0$; similarly, saying that the system (2.5.4) has q linearly independent solutions $(c_{1r}, c_{2r}, \ldots, c_{nr})$ is equivalent to saying that the problem (2.5.6) has q linearly independent solutions

$$U_r(x) = \sum_{j=1}^{n} c_{jr}\, u_j(x)\,, \qquad (r = 1, 2, \ldots, q)\,. \tag{2.5.7}$$

In the latter case, the necessary and sufficient conditions (2.5.5) that the quadrature formula (2.5.2) exist, can be written as

$$\int_a^b g(x)\, U_r(x)\, dx = 0\,, \qquad (r = 1, 2, \ldots, q)\,. \tag{2.5.8}$$

Therefore we may state the following theorem:

Theorem 2.5.I. *Given the nodes $x_1, x_2, \ldots, x_m$, the linear differential operator E of order n and the positive integer $p \leqq n - 1$, consider the homogeneous differential problem (2.5.6). If this problem has no non-trivial solutions [whence $n \leqq m\,(n - p)$] it is possible to write a quadrature formula of the type (2.5.2) in $\infty^{m(n-p)-n}$ different ways. If on the other hand the problem (2.5.6) has q linearly independent solutions $U_r(x)$ [$r = 1, 2, \ldots, q$ with $n - m\,(n - p) \leqq q \leqq p$] then the formula (2.5.2) may apply only if the q conditions (2.5.8) are satisfied; if so there are $\infty^{m(n-p)-n+q}$ possible formulae of form (2.5.2).*

In the conditions stated in this theorem, the equation (2.5.8) may appear as conditions for the weight $g(x)$ (see § 2.7, Problem 4); but this point of view has little importance since generally the weight $g(x)$ is given.

However we may fix only the number m of the nodes and consider their abscissae $x_1, x_2, \ldots, x_m$, all or in part, as undetermined parameters. In this case, (2.5.8) appear as *conditions for the nodes*; they show us how we have to choose these points to make it possible to get the quadrature formula (2.5.2) required.[1]) This problem, presented in these terms, will be called *Gauss problem* since it is a wide generalization of a particular problem studied by Gauss which originated the quadrature formulae having his name (see § 4.6).

2.6. Tchebychef problem.[2]) Now let us consider another problem regarding quadrature formulae which, in certain particular cases, have been studied by Tchebychef. In analogy with Gauss problem [see (2.5.2)] we try to

[1]) It may be that there many groups of nodes $(x_1, \ldots, x_m)$ satisfying (2.5.8). In that case it is obvious that the latter statement of theorem 2.5.I (existence of $\infty^{m(n-p)-n+q}$ quadrature formulae) is to be applied to each of such groups (see § 2.7, Problems 2, 3).

[2]) See A. Ghizzetti [1].

get, for a fixed p $(0 \leqq p \leqq n-1)$, a quadrature formula of the type

$$\int_a^b g(x)\, u(x)\, dx = \sum_{h=0}^{n-p-1} A_h \sum_{i=1}^{m} u^{(h)}(x_i) + R(u)\,, \qquad [E(u) = 0 \Rightarrow R(u) = 0]\,, \tag{2.6.1}$$

that is, one in which *the coefficients of the values* $u^{(h)}(x_i)$, *of the derivatives of the same order of the function* $u(x)$, *are always equal to each other.* Reasoning as in the preceding § 2.5 we find for the $n-p$ coefficients A_h the following system of n linear equations

$$\sum_{h=0}^{n-p-1} A_h \sum_{i=1}^{m} u_j^{(h)}(x_i) = \int_a^b g(x)\, u_j(x)\, dx\,, \qquad (j = 1, 2, \ldots, n)\,, \tag{2.6.2}$$

whose homogeneous transpose can be written

$$\sum_{j=1}^{n} c_j \sum_{i=1}^{m} u_j^{(h)}(x_i) = 0\,, \qquad (h = 0, 1, \ldots, n-p-1)\,. \tag{2.6.3}$$

If the rank of the matrix $\left(\sum_{i=1}^{m} u_j^{(h)}(x_i)\right)$ (with $n-p$ rows and n columns) comes out to be equal to n (this is possible only if $p = 0$), the system (2.6.3) has no non-trivial solutions, therefore the system (2.6.2) is consistent and admits of one and only one solution. Instead let us assume that the matrix has rank $n-q$ (with $q \geqq 1$, $0 \leqq n-q \leqq n-p$). In this case the system (2.6.3) has q linearly independent solutions $(c_{1r}, c_{2r}, \ldots, c_{nr})$ $(r = 1, 2, \ldots, q)$ and the system (2.6.2) is consistent if and only if the q conditions

$$\sum_{j=1}^{n} c_{jr} \int_a^b g(x)\, u_j(x)\, dx = 0\,, \qquad (r = 1, 2, \ldots, q)\,, \tag{2.6.4}$$

are satisfied, resulting $\infty^{(n-p)-(n-q)}$ solutions of (2.6.1). These results can be presented in a different way, considering the following homogeneous boundary problem:

$$E(u) = 0; \qquad \sum_{i=1}^{m} u^{(h)}(x_i) = 0\,, \qquad (h = 0, 1, \ldots, n-p-1)\,. \tag{2.6.5}$$

Since the general solution of $E(u) = 0$ is $u(x) = \sum_{j=1}^{n} c_j\, u_j(x)$ assuming the boundary conditions satisfied, we get for the arbitrary constants c_j precisely the system (2.6.3). Therefore when we say that (2.6.3) has no non-trivial solutions we mean that (2.6.5) admits only the solution $u(x) = 0$; to say that (2.6.3) has q linearly independent solutions $(c_{1r}, c_{2r}, \ldots, c_{nr})$ means that the problem (2.6.5) has q linearly independent solutions

$$U_r(x) = \sum_{j=1}^{n} c_{jr}\, u_j(x)\,, \qquad (r = 1, 2, \ldots, q)\,, \tag{2.6.6}$$

and we write (2.6.4) as

$$\int_a^b g(x)\, U_r(x)\, dx = 0\,, \qquad (r = 1, 2, \ldots, q)\,. \tag{2.6.7}$$

We have then proved the following theorem:

Theorem 2.6.I. *Given the nodes $x_1, x_2, \ldots, x_m$, the linear differential operator E of order n and the non-negative integer $p \leqq n-1$, if the problem (2.6.5) has no non-trivial solutions (this is possible only if $p = 0$) we may get a quadrature formula of the type (2.6.1) in only one way.*

If on the other hand problem (2.6.5) has q linearly independent solutions $U_r(x)$, $(r = 1, 2, \ldots, q$ with $q \geqq 1$, $p \leqq q \leqq n)$ [1]) *then it is possible to construct a formula (2.6.1) (in ∞^{q-p} different ways) if and only if the q conditions (2.6.7) are satisfied.*

We may repeat here the remarks stated in the preceding § 2.5: (2.6.7) may be considered as conditions on the weight g (but this has a little importance) or as *conditions on the choice of the nodes.*

2.7. Problems.

1. Examine the possibility of devise a quadrature formula of the type

$$\int_a^b u(x)\,dx = A_{01}\,u(x_1) + A_{02}\,u(x_2) + A_{03}\,u(x_3) + R(u) \tag{1}$$

[with $a \leqq x_1 < x_2 < x_3 \leqq b$] which must be exact when $u(x)$ is a polynomial of degree $\leqq 3$.

2. The Gauss problem of § 2.5 can be generalized by substituting (2.5.2) by the formula

$$\int_a^b g(x)\,u(x)\,dx = \sum_{i=1}^{m}\sum_{h=0}^{n-p_i-1} A_{hi}\,u^{(h)}(x_i) + R(u)\,, \tag{1}$$

$$E(u) = 0 \Rightarrow R(u) = 0\,,$$

where the maximum order $n - p_i - 1$ of the derivatives of the argument function $u(x)$, which must be considered in the node x_i, depends on the node x_i. Prove the following theorem (which is an extension of theor. 2.5.I): *given the nodes $x_1, x_2, \ldots, x_m$, the linear diffential operator E of order n and the integers $p_1, p_2, \ldots, p_m$ (with $0 \leqq p_i \leqq n-1$), consider the boundary problem:*

$$E(u) = 0;\; u^{(h)}(x_i) = 0,\; (i = 1, 2, \ldots, m;\; h = 0, 1, \ldots, n - p_i - 1)\,. \tag{2}$$

If this problem has only the trivial solution $u(x) \equiv 0$ [whence $n \leqq m - \sum_{i=1}^{m} p_i$] it is possible to write a quadrature formula of the type (1) in $\infty^{mn - \sum_{i=1}^{m} p_i - n}$

[1]) It may happen that $q = n$, namely that all the solutions of $E(u) = 0$ satisfy the boundary conditions $\sum_{i=1}^{m} u^{(h)}(x_i) = 0$, $(h = 0, 1, \ldots, n - p - 1)$. Example: $n = 2$, $m = 2$, $p = 0$, $E = \frac{d^2}{dx^2} + 1$, $x_1 = 0$, $x_2 = \pi$.

different ways. If on the other hand the problem (2) *has q linearly independent solutions* $U_r(x)$ $[r = 1, 2, \ldots, q$ *with* $n - m\,n + \sum_{i=1}^{m} p_i \leqq q \leqq \min p_i]$, *then the formula* (1) *is possible if and only if these q conditions are satisfied:*

$$\int_a^b g(x)\, U_r(x)\, dx = 0\,, \qquad (r = 1, 2, \ldots, q); \tag{3}$$

in such a case this formula is possible in $\infty^{mn - \sum_{i=1}^{m} p_i - n + q}$ *different ways.*

3. Examine the possibility of devise a quadrature formula of the type

$$\int_{-1}^{1} u(x)\, dx = A_{01}\, u(x_1) + A_{11}\, u'(x_1) + A_{21}\, u''(x_1) + A_{02}\, U(x_2) + R(u)\,, \quad (1)$$

which must be exact when $u(x)$ is a polynomial of degree $\leqq 5$.

4. Determine the weight functions $g(x)$ in correspondence of which it is possible to devise a quadrature formula of the type

$$\int_{-1}^{1} g(x)\, u(x)\, dx = A_{01}\, u(-1) + A_{02}\, u(0) + A_{03}\, u(1) + R(u)\,, \quad (1)$$

which must be exact when $u(x)$ is a polynomial of degree $\leqq 5$.

CHAPTER 3

Special functions

3.1. Bernouilli polynomials and numbers. The *Bernouilli polynomials* $B_n(x)$ are defined by the following expansion (valid for $|z| < 2\pi$):

$$\left.\begin{array}{c} \dfrac{z\, e^{xz}}{e^z - 1} = \sum_{n=0}^{\infty} B_n(x) \dfrac{z^n}{n!}\,, \text{[1]} \\ \left[B_0(x) = 1\,, \quad B_1(x) = x - \dfrac{1}{2}\,, \quad B_2(x) = x^2 - x + \dfrac{1}{6}\,, \ldots,\right] \end{array}\right\} \tag{3.1.1}$$

while the *Bernouilli numbers* B_n are given by

$$B_n = B_n(0)\,, \quad \left[B_0 = 1\,, \quad B_1 = -\frac{1}{2}\,, \quad B_2 = \frac{1}{6}\,, \ldots\right]; \tag{3.1.2}$$

therefore we have

$$\frac{z}{e^z - 1} = \sum_{n=0}^{\infty} B_n \frac{z^n}{n!} = 1 - \frac{1}{2} z + \sum_{n=2}^{\infty} B_n \frac{z^n}{n!} \tag{3.1.3}$$

and since $\dfrac{z}{e^z - 1} - 1 + \dfrac{1}{2} z = \dfrac{z \cosh z/2}{2 \sinh z/2} - 1$ is an even function, we deduce that

$$B_3 = B_5 = B_7 = \cdots = 0\,, \tag{3.1.4}$$

while the other Bernouilli numbers $B_2, B_4, B_6, \ldots$ are all non zero.

Now note that from the identity

$$z\, e^{xz} = \frac{z\, e^{(x+1)z}}{e^z - 1} - \frac{z\, e^{xz}}{e^z - 1}$$

we may deduce, expanding every term in power series of z:

$$\sum_{n=0}^{\infty} x^n \frac{z^{n+1}}{n!} = \sum_{n=0}^{\infty} B_n(x+1) \frac{z^n}{n!} - \sum_{n=0}^{\infty} B_n(x) \frac{z^n}{n!}$$

or also

$$\sum_{n=1}^{\infty} n\, x^{n-1} \frac{z^n}{n!} = \sum_{n=1}^{\infty} [B_n(x+1) - B_n(x)] \frac{z^n}{n!}\,;$$

therefore we have

$$B_n(x+1) - B_n(x) = n\, x^{n-1}\,, \quad (n = 1, 2, \ldots) \tag{3.1.5}$$

and particularly, putting $x = 0$ and using (3.1.2)

$$B_n(1) = B_n\,, \qquad (n = 2, 3, \ldots)\,. \tag{3.1.6}$$

[1]) That $B_n(x)$ come out to be effectively a polynomial of degree n, it immediately follows from the formulae given furtherwards [for instance from (3.1.8)].

From equation (3.1.1) there follows, differentiating with respect to x

$$\frac{z^2\, e^{xz}}{e^z - 1} = \sum_{n=1}^{\infty} B_n'(x)\, \frac{z^n}{n!}\,; \tag{3.1.7}$$

on the other part, always for (3.1.1), the function occurring in the first member has value $\sum_{n=0}^{\infty} B_n(x)\, \frac{z^{n+1}}{n!} = \sum_{n=1}^{\infty} n\, B_{n-1}(x)\, \frac{z^n}{n!}$, so that from the comparison with (3.1.7) we deduce

$$B_n'(x) = n\, B_{n-1}(x)\,, \qquad (n = 1, 2, \ldots)\,. \tag{3.1.8}$$

From this equation with further differentiations we obtain $B_n''(x) = n \times B_{n-1}'(x) = n\,(n-1)\, B_{n-2}(x)$, $B_n'''(x) = n\,(n-1)\, B_{n-2}'(x) = n\,(n-1)\,(n-2) \times B_{n-3}(x)$, ... and in general

$$B_n^{(k)}(x) = \frac{n!}{(n-k)!}\, B_{n-k}(x)\,, \qquad (k = 0, 1, 2, \ldots, n)\,. \tag{3.1.9}$$

Supposing $k \leqq n - 2$, that is $n - k \geqq 2$, if we put $x = 0$ or $x = 1$ in (3.1.9), using (3.1.2) and (3.1.6), we obtain

$$B_n^{(k)}(0) = B_n^{(k)}(1) = \frac{n!}{(n-k)!}\, B_{n-k}\,, \qquad (k = 0, 1, \ldots, n-2)\,, \tag{3.1.10}$$

while in the case of $k = n - 1$, since (3.1.9) becomes $B_n^{(n-1)}(x) = n!\left(x - \frac{1}{2}\right)$, we obtain

$$B_n^{(n-1)}(0) = -\frac{1}{2}\, n!\,, \qquad B_n^{(n-1)}(1) = \frac{1}{2}\, n!\,. \tag{3.1.11}$$

We get another interesting consequence of (3.1.8) integrating it between 0 and x; then we have

$$B_n(x) - B_n = n \int_0^x B_{n-1}(\xi)\, d\xi\,, \qquad (n = 1, 2, \ldots)\,. \tag{3.1.12}$$

From it, changing n by $n + 1$ and putting $x = 1$, using (3.1.6) and supposing $n \geqq 1$ we get:

$$\int_0^1 B_n(x)\, dx = 0\,, \qquad (n = 1, 2, \ldots)\,. \tag{3.1.13}$$

Let us now deduce a property of symmetry or semisymmetry of the polynomials $B_n(x)$ with respect to the point $x = \frac{1}{2}$. We write again (3.1.1), a first time changing x by $1 - x$ and a second time changing z by $-z$, thus obtaining the two formulae

$$\frac{z\, e^{(1-x)z}}{e^z - 1} = \sum_{n=0}^{\infty} B_n(1-x)\, \frac{z^n}{n!}\,, \qquad -\frac{z\, e^{-xz}}{e^{-z} - 1} = \sum_{n=0}^{\infty} (-1)^n\, B_n(x)\, \frac{z^n}{n!}\,;$$

it is easy to see that the first members are equal to each other and therefore we have the formula

$$B_n(1-x) = (-1)^n\, B_n(x)\,, \qquad (n = 0, 1, 2, \ldots) \tag{3.1.14}$$

which expresses the wanted property.

Finally let us consider the polynomial

$$P_{2n}(x) = B_{2n}(x) - B_{2n}, \quad (n = 1, 2, \ldots), \tag{3.1.15}$$

on the interval [0, 1]. Using (3.1.2), (3.1.4), (3.1.6), (3.1.10) we see that

$$\left.\begin{aligned} &P_{2n}(0) = P_{2n}(1) = 0, \\ &P_{2n}^{(2k-1)}(0) = P_{2n}^{(2k-1)}(1) = 0, \qquad (k = 1, 2, \ldots, n-1), \\ &P_{2n}^{(2k)}(0) = P_{2n}^{(2k)}(1) \neq 0, \qquad (k = 1, 2, \ldots, n); \end{aligned}\right\} \tag{3.1.16}$$

wherefrom and from the Rolle theorem it is easy to deduce that $P_{2n}(x)$ *does not vanish for* $0 < x < 1$. In fact if $P_{2n}(x)$ vanishes (besides at the points $x = 0$, $x = 1$ also) at a point internal in [0, 1], its derivative $P'_{2n}(x)$ should vanish (besides at the points $x = 0$, $x = 1$) at least at two internal points, $P''_{2n}(x)$ at least at three internal points, $P'''_{2n}(x)$ (besides at $x = 0$, $x = 1$) at least at two internal points and so on until the absurd conclusion that $P_{2n}^{(2n-2)}(x)$ (polynomial of second degree) should vanish at least at three points internal in [0, 1].

This reasoning proves also that $P'_{2n}(x)$ vanishes only at one point internal in [0, 1], which point has necessarily abscissa $\frac{1}{2}$ in virtue of the symmetry property expressed by (3.1.14). Therefore $P_{2n}(x)$ *is always positive* (*or negative*) *for* $0 < x < 1$ *and gets its maximum* (*or minimum*) *value for* $x = \frac{1}{2}$.

For computing $P_n\left(\frac{1}{2}\right)$ we write (3.1.1) with $x = 1$:

$$\frac{z\,e^z}{e^z - 1} = \sum_{n=0}^{\infty} B_n(1)\,\frac{z^n}{n!}; \tag{3.1.17}$$

we observe then that the first member may also be written

$$\frac{1}{2}\left(\frac{2\,z\,e^{\frac{1}{2}2z}}{e^{2z} - 1} + \frac{2\,z\,e^{2z}}{e^{2z} - 1}\right)$$

so that, applying again (3.1.1) with $x = \frac{1}{2}$ or $x = 1$ and $2\,z$ instead of z, we obtain

$$\frac{z\,e^z}{e^z - 1} = \frac{1}{2}\left[\sum_{n=0}^{\infty} B_n\left(\frac{1}{2}\right)\frac{(2\,z)^n}{n!} + \sum_{n=0}^{\infty} B_n(1)\,\frac{(2\,z)^n}{n!}\right]. \tag{3.1.18}$$

From the comparison between (3.1.17) and (3.1.18) obviously follows that $B_n(1) = 2^{n-1}\left[B_n\left(\frac{1}{2}\right) + B_n(1)\right]$, whence we have $B_n\left(\frac{1}{2}\right) = -\left(1 - \frac{1}{2^{n-1}}\right) \times B_n(1)$ and finally, using (3.1.6):

$$B_n\left(\frac{1}{2}\right) = -\left(1 - \frac{1}{2^{n-1}}\right) B_n, \quad (n = 2, 3, \ldots). \tag{3.1.19}$$

From equations (3.1.15) and (3.1.19) then follows that *the above mentioned maximum value* (*or minimum*) *of* $P_{2n}(x)$ *in* [0, 1] *is given by*

$$P_{2n}\left(\frac{1}{2}\right) = -2\left(1 - \frac{1}{2^{2n}}\right) B_{2n}, \quad (n = 1, 2, \ldots). \tag{3.1.20}$$

We may speak of maximum if $B_{2n} < 0$, of minimum if $B_{2n} > 0$. In the first case the polynomial $P_{2n}(x)$ has in $[0, 1]$ the minimum (equal to zero) at the extreme points $x = 0$ and $x = 1$; from it and from the fact for (3.1.10) we have $P''_{2n}(0) = P''_{2n}(1) = 2n(2n-1)B_{2n-2}$ follows $B_{2n-2} > 0$. Analogously in the other case we obtain $B_{2n-2} < 0$. We conclude that B_{2n-2} and B_{2n} have always opposite signs and therefore, being $B_2 > 0$, that B_{2n} has the sign of $(-1)^{n-1}$; there follows that $P_{2n}(x)$ *has for* $0 < x < 1$ *the sign of* $(-1)^n$.

3.2. Euler functions. *The Euler function of second kind* or *gamma function* is defined for $\alpha > 0$ by the following formula

$$\Gamma(\alpha) = \int_0^{+\infty} e^{-x} x^{\alpha-1}\, dx\,, \qquad (\alpha > 0)\,, \tag{3.2.1}$$

from which we immediately deduce

$$\Gamma(1) = 1\,, \tag{3.2.2}$$

$$\Gamma\left(\frac{1}{2}\right) = \int_0^{+\infty} \frac{e^{-x}}{\sqrt{x}}\, dx = 2\int_0^{+\infty} e^{-t^2}\, dt = \sqrt{\pi}\,. \tag{3.2.3}$$

Then we have

$$\Gamma(\alpha+1) = \int_0^{+\infty} e^{-x} x^{\alpha}\, dx = [-x^{\alpha} e^{-x}]_0^{+\infty} + \alpha\int_0^{+\infty} e^{-x} x^{\alpha-1}\, dx$$

that is

$$\Gamma(\alpha+1) = \alpha\,\Gamma(\alpha)\,. \tag{3.2.4}$$

For n positive integer, a repeated application of this formula yields

$$\Gamma(\alpha+n) = (\alpha+n-1)(\alpha+n-2)\cdots(\alpha+1)\,\alpha\,\Gamma(\alpha)\,, \tag{3.2.5}$$

and particularly putting $\alpha = 1$ or $\alpha = \frac{1}{2}$ and using (3.2.2) and (3.2.3):

$$\Gamma(n+1) = n! \tag{3.2.6}$$

$$\Gamma\left(n+\frac{1}{2}\right) = \left(n-\frac{1}{2}\right)\left(n-\frac{3}{2}\right)\cdots\frac{3}{2}\cdot\frac{1}{2}\sqrt{\pi} = \frac{(2n)!}{2^{2n}\, n!}\sqrt{\pi}\,. \tag{3.2.7}$$

From equation (3.2.4) follows too

$$\Gamma(\alpha+1) = \alpha(\alpha-1)(\alpha-2)\cdots(\alpha-n+1)\,\Gamma(\alpha-n+1)\,, \quad (\alpha > n-1)$$

whence the binomial coefficients $\binom{\alpha}{n}$ may be expressed as follows

$$\binom{\alpha}{n} = \frac{1}{n!}\,\frac{\Gamma(\alpha+1)}{\Gamma(\alpha-n+1)}\,, \qquad (\alpha > n-1)\,. \tag{3.2.8}$$

The *Euler function of first kind* or *beta function* is defined by the formula

$$B(\alpha, \beta) = \int_0^1 x^{\alpha-1}(1-x)^{\beta-1}\, dx\,, \qquad (\alpha > 0,\ \beta > 0)\,. \tag{3.2.9}$$

Now we prove that this function is connected to the gamma function by the following relation

$$B(\alpha, \beta) = \frac{\Gamma(\alpha)\,\Gamma(\beta)}{\Gamma(\alpha + \beta)}\,. \tag{3.2.10}$$

In fact for (3.2.1) we have

$$\Gamma(\alpha)\,\Gamma(\beta) = \int_0^{+\infty} e^{-x}\, x^{\alpha-1}\, dx \cdot \int_0^{+\infty} e^{-y}\, y^{\beta-1}\, dy = = \iint_Q e^{-(x+y)}\, x^{\alpha-1}\, y^{\beta-1}\, dx\, dy\,,$$

where Q denotes the quadrant $x \geqq 0$, $y \geqq 0$ of the plane $x\,y$. Operating in this double integral the changing of variables $x = u\,v$, $y = (1 - u)\,v$, which transforms the quadrant Q into the strip S defined by $0 \leqq u \leqq 1$, $v \geqq 0$, we get

$$\begin{aligned}\Gamma(\alpha)\,\Gamma(\beta) &= \iint_S e^{-v}\, u^{\alpha-1}\, v^{\alpha+\beta-1}\,(1 - u)^{\beta-1}\, du\, dv \\ &= \int_0^1 u^{\alpha-1}\,(1 - u)^{\beta-1}\, du \cdot \int_0^{+\infty} e^{-v}\, v^{\alpha+\beta-1}\, dv;\end{aligned}$$

but for (3.2.1) and (3.2.9) the latter expression has value $B(\alpha, \beta)\,\Gamma(\alpha + \beta)$ and (3.2.10) follows.

Finally we prove, for the gamma function, the following *Legendre duplication formula*

$$\Gamma(2\,\alpha) = \frac{1}{\sqrt{\pi}}\, 2^{2\alpha-1}\, \Gamma(\alpha)\, \Gamma\left(\alpha + \frac{1}{2}\right). \tag{3.2.11}$$

From (3.2.9) follows

$$B(\alpha, \alpha) = \int_0^1 x^{\alpha-1}\,(1 - x)^{\alpha-1}\, dx = 2\int_0^{1/2} (x - x^2)^{\alpha-1}\, dx$$

and operating on this integral the substitution $x = (1 - \sqrt{1 - y})/2$, which implies $y = 4\,(x - x^2)$:

$$B(\alpha, \alpha) = 2\int_0^1 \left(\frac{y}{4}\right)^{\alpha-1} \frac{dy}{4\sqrt{1 - y}} = \frac{1}{2^{2\alpha-1}} \int_0^1 y^{\alpha-1}\,(1 - y)^{-\frac{1}{2}}\, dy\,,$$

that is for (3.2.9)

$$B(\alpha, \alpha) = \frac{1}{2^{2\alpha-1}}\, B\left(\alpha, \frac{1}{2}\right).$$

For (3.2.10) this formula may be transformed into

$$\frac{[\Gamma(\alpha)]^2}{\Gamma(2\,\alpha)} = \frac{1}{2^{2\alpha-1}}\, \frac{\Gamma(\alpha)\, \Gamma\left(\frac{1}{2}\right)}{\Gamma\left(\alpha + \frac{1}{2}\right)}$$

which, using (3.2.3), is equivalent to (3.2.11).

3.3. Orthogonal polynomials connected to a given weight function. Let a weight function $p(x)$, measurable and non negative, be defined on an interval $[a, b]$, finite or infinite, being $p(x) > 0$ in a set of positive measure and $p(x)\, x^n \in L[a, b]$ for $n = 0, 1, 2, \ldots$;[1]) therefore there exist finite all the *moments*

$$\mu_n = \int_a^b p(x)\, x^n\, dx\,, \qquad (n = 0, 1, 2, \ldots)\,, \tag{3.3.1}$$

of $p(x)$ and we have $\mu_0 > 0$. Let then a sequence $\{a_n\}_{n=0,1,2,\ldots}$ of numbers *all different from zero* be assigned.

Then the following theorem holds:

Theorem 3.3.I. *It is possible to construct in one and only one way a sequence* $\{P_n(x)\}_{n=0,1,2,\ldots}$ *with* $P_n(x)$ *polynomial of n degree having* a_n *as coefficients of* x^n, *in such a way that*

$$\int_a^b p(x)\, P_m(x)\, P_n(x)\, dx = 0 \quad \text{if} \quad m \neq n\,, \tag{3.3.2}$$

or, under an equivalent form

$$\int_a^b p(x)\, \Pi_{n-1}(x)\, P_n(x)\, dx = 0\,, \qquad (n = 1, 2, \ldots)\,, \tag{3.3.3}$$

where $\Pi_{n-1}(x)$ *denotes an arbitrary polynomial of degree* $\leqq n - 1$.

Proof. For the polynomial $P_0(x)$ we surely have $P_0(x) = a_0$, while the next polynomial $P_1(x) = a_1 x + b_1$ is determined by $\int_a^b p(x)\, P_0(x)\, P_1(x)\, dx = a_0 (a_1 \mu_1 + b_1 \mu_0) = 0$ which yields $b_1 = -\, a_1 \frac{\mu_1}{\mu_0}$. We have only to prove that, supposing $P_0(x), P_1(x), \ldots, P_n(x)$ with $n \geqq 1$ already constructed, $P_{n+1}(x) = a_{n+1}\, x^{n+1} + \cdots$ is determined.

We evidently may seek the polynomial $P_{n+1}(x)$ with the form

$$P_{n+1}(x) = \frac{a_{n+1}}{a_n}\, x\, P_n(x) + \sum_{k=1}^{n+1} c_k\, P_{n+1-k}(x) \tag{3.3.4}$$

where the coefficients $c_1, c_2, \ldots, c_{n+1}$ are unknown. Having fixed m (with $m = 0, 1, \ldots, n$), we multiply both members of (3.3.4) by $p(x)\, P_m(x)$ and then we integrate on $[a, b]$; this leads to

$$\int_a^b p(x)\, P_m(x)\, P_{n+1}(x)\, dx = \frac{a_{n+1}}{a_n} \int_a^b p(x)\, x\, P_m(x)\, P_n(x)\, dx$$

$$+ \sum_{k=1}^{n+1} c_k \int_a^b p(x)\, P_m(x)\, P_{n+1-k}(x)\, dx\,, \qquad (m = 0, 1, \ldots, n)\,.$$

[1]) Of course, if $[a, b]$ is finite, these hypotheses reduce to the one $p(x) \in L[a, b]$.

In virtue of (3.3.2) [or (3.3.3)] this relation yields

$$\left.\begin{aligned}
0 &= c_{n+1-m}\int_a^b p(x)\,P_m^2(x)\,dx\,, \qquad (m = 0, 1, \ldots, n-2)\,,\\
0 &= \frac{a_{n+1}}{a_n}\int_a^b p(x)\,x\,P_{n-1}(x)\,P_n(x)\,dx + c_2\int_a^b p(x)\,P_{n-1}^2(x)\,dx\,,\\
0 &= \frac{a_{n+1}}{a_n}\int_a^b p(x)\,x\,P_n^2(x)\,dx + c_1\int_a^b p(x)\,P_n^2(x)\,dx\,,
\end{aligned}\right\} \tag{3.3.5}$$

whence (3.3.4) becomes

$$P_{n+1}(x) = \frac{a_{n+1}}{a_n}\,x\,P_n(x) + c_1\,P_n(x) + c_2\,P_{n-1}(x)\,, \tag{3.3.6}$$

where the constants c_1, c_2 are given by the second and the third equation of (3.3.5). Equation (3.3.6) proves that $P_{n+1}(x)$ is completely determined; vice versa it is soon proved that this $P_{n+1}(x)$ satisfies $\int_a^b p\,P_{n+1}\,P_k\,dx = 0$, $(k = 0, 1, \ldots, n)$. The theorem is then proved.

The preceding proof also proves that we have

$$P_n(x) = a_n\,P_n^*(x)\,, \qquad (n = 0, 1, 2, \ldots)\,, \tag{3.3.7}$$

where $P_n^*(x)$ *is a polynomial fully determined by the weight* $p(x)$ (that one which corresponds to $a_n = 1$).

We say that a sequence of polynomials $P_n(x)$ satisfying (3.3.2) *forms a system of orthogonal polynomials with respect to the weight* $p(x)$. On the basis of (3.3.7) it is clear that the *only conditions (3.3.2) determine every polynomial* $P_n(x)$ *only apart from an arbitrary constant factor* $a_n \neq 0$.

This arbitrary factor can be fixed with different criteria; one of the most common is that of rendering the polynomials *orthogonal and normal* (or briefly *orthonormal*), that is such as to give

$$\int_a^b p(x)\,P_m(x)\,P_n(x)\,dx = \delta_{mn}\,. \tag{3.3.8}$$

It is clear that, to this purpose, it suffices to take the factors a_n in such a way that $a_n^2\int_a^b p(x)\,P_n^{*2}(x)\,dx = 1$, that is to put

$$P_n(x) = \frac{P_n^*(x)}{\sqrt{\int_a^b p(x)\,P_n^{*2}(x)\,dx}}\,, \qquad (n = 0, 1, 2, \ldots)\,. \tag{3.3.9}$$

We now observe that equation (3.3.6) shows that among three consecutive polynomials of our orthogonal system there exists a *recurrence formula* of the type

$$x P_n(x) = A_n P_{n+1}(x) + B_n P_n(x) + C_n P_{n-1}(x), \qquad (3.3.10)$$
$$(n = 0, 1, 2, \ldots), ^{1)}$$

where the coefficients A_n, B_n, C_n have values easily deducible from (3.3.5) and (3.3.6). We find

$$A_n = \frac{a_n}{a_{n+1}}, \qquad (3.3.11)$$

$$B_n = -\frac{a_n}{a_{n+1}} c_1 = \frac{\int_a^b p(x)\, x\, P_n^2(x)\, dx}{\int_a^b p(x)\, P_n^2(x)\, dx}, \qquad (3.3.12)$$

$$C_n = -\frac{a_n}{a_{n+1}} c_2 = \frac{\int_a^b p(x)\, x\, P_{n-1}(x)\, P_n(x)\, dx}{\int_a^b p(x)\, P_{n-1}^2(x)\, dx} = \frac{a_{n-1}}{a_n} \frac{\int_a^b p(x)\, P_n^2(x)\, dx}{\int_a^b p(x)\, P_{n-1}^2(x)\, dx}. ^{2)} \qquad (3.3.13)$$

Another expression of B_n can be obtained by putting $P_n(x) = a_n x^n + b_n x^{n-1} + \cdots$, substituting in (3.3.10) and equalizing the coefficients of x^{n+1} and x^n in both members. We find $a_n = A_n a_{n+1}$, $b_n = A_n b_{n+1} + B_n a_n$ from which there follows (3.3.11) and

$$B_n = \frac{b_n}{a_n} - \frac{b_{n+1}}{a_{n+1}}. \qquad (3.3.14)$$

Other properties of the orthogonal polynomials $P_n(x)$ are expressed by the following theorems:

Theorem 3.3.II. *Supposing orthonormal the system* $\{P_n(x) = a_n x^n + \cdots\}$; *there exists the following Christoffel-Darboux identity:*

$$\sum_{k=0}^{n} P_k(x)\, P_k(y) = \frac{a_n}{a_{n+1}} \frac{P_{n+1}(x)\, P_n(y) - P_{n+1}(y)\, P_n(x)}{x - y}, \qquad (n = 0, 1, 2, \ldots). \qquad (3.3.15)$$

Proof. We observe that, being the system orthonormal, (3.3.13) becomes $C_n = \frac{a_{n-1}}{a_n}$ whence, considering again (3.3.10) and multiplying its both

[1]) This formula is valid also for $n = 0$, whether we agree that $P_{-1}(x) = 0$.

[2]) The last passage can be justified thinking that $x P_{n-1}(x)$ is the sum of $\frac{a_{n-1}}{a_n} P_n(x)$ and of a linear combination of $P_{n-1}(x), \ldots, P_0(x)$; then for the orthogonality conditions we have

$$\int_a^b p\, x\, P_{n-1}\, P_n\, dx = \frac{a_{n-1}}{a_n} \int_a^b p\, P_n^2\, dx.$$

members by $P_n(y)$, we may write (changing n by k):

$$x\, P_k(x)\, P_k(y) = \frac{a_k}{a_{k+1}} P_{k+1}(x)\, P_k(y) + B_k\, P_k(x)\, P_k(y) + \frac{a_{k-1}}{a_k} P_{k-1}(x)\, P_k(y)\,.$$

Subtrating from it the analogous relation deduced by changing x by y, we obtain

$$(x-y)\, P_k(x)\, P_k(y) = \frac{a_k}{a_{k+1}} [P_{k+1}(x)\, P_k(y) - P_{k+1}(y)\, P_k(x)]$$
$$- \frac{a_{k-1}}{a_k} [P_k(x)\, P_{k-1}(y) - P_k(y)\, P_{k-1}(x)]\,.$$

Summing with respect to $k = 0, 1, \ldots, n$ and recalling that $P_{-1}(x) = 0$ we obtain

$$(x-y) \sum_{k=0}^{n} P_k(x)\, P_k(y) = \frac{a_n}{a_{n+1}} [P_{n+1}(x)\, P_n(y) - P_{n+1}(y)\, P_n(x)]$$

which is equivalent to (3.3.15) q.e.d.

Theorem 3.3.III. *The n zeros of the polynomial $P_n(x)$, $(n = 1, 2, \ldots)$ are real, distinct and internal in the interval $[a, b]$.*

Proof. Let us denote by $x_1, x_2, \ldots, x_m$ (with $0 \leqq m \leqq n$) the possible real zeros of odd multiplicity of $P_n(x)$, internal in the interval $[a, b]$. The theorem will be proved by showing that necessarily $m = n$.

We consider the following polynomial of degree m

$$\Pi_m(x) = \begin{cases} 1 & (\text{if } m = 0)\,, \\ \prod_{i=1}^{m} (x - x_i) & (\text{if } m = 1, 2, \ldots, n)\,. \end{cases}$$

In the interval $[a, b]$ this polynomial changes its sign in the same points where the polynomial $P_n(x)$ changes its own and therefore the product $\Pi_m(x)\, P_n(x)$ does not change sign.

On the basis of the hypotheses done on the weight $p(x)$ there follows

$$\int_a^b p(x)\, \Pi_m(x)\, P_n(x)\, dx \neq 0\,,$$

but for (3.3.3) it cannot exist if $m < n$; then necessarily it is $m = n$, q.e.d.

Theorem 3.3.IV. *Let the interval $[a, b]$ be of the type $[-c, c]$ and the weight $p(x)$ be an even function. Then the polynomial $P_n(x)$ is an even function if n is even, odd function if n is odd.*

Proof. Let us consider the following integral

$$I_n = \int_{-c}^{c} p(x)\, \Pi_{n-1}(x)\, P_n(-x)\, dx\,, \quad (n = 1, 2, \ldots)\,, \qquad (3.3.16)$$

where $\Pi_{n-1}(x)$ denotes any polynomial of degree $\leqq n - 1$. Operating on this integral the substitution $x = -t$ we obtain

$$I_n = \int_{-c}^{c} p(t)\, \Pi_{n-1}(-t)\, P_n(t)\, dt$$

whence recalling (3.3.3) we have $I_n = 0$. From (3.3.16) then follows that $P_n(-x)$ differs from $P_n(x)$ for a constant factor, which is necessarily equal to $(-1)^n$; we have then $P_n(-x) = (-1)^n P_n(x)$ and this proves the theorem.

Let us now consider attentively these three particular cases:

$$a, b \text{ finite}\ ;\quad p(x) = (b - x)^\alpha (x - a)^\beta\ ;\quad \alpha > -1\,,\quad \beta > -1\,, \tag{3.3.17}$$

$$a \text{ finite}\,,\quad b = +\infty\ ;\quad p(x) = (x - a)^\alpha e^{-kx}\ ;\quad \alpha > -1\,,\quad k > 0\,, \tag{3.3.18}$$

$$a = -\infty\,,\quad b = +\infty\ ;\quad p(x) = e^{hx - kx^2}\ ;\quad k > 0\,, \tag{3.3.19}$$

in which the weight $p(x)$ can be characterized (apart from constant factor) by the fact that it is solution, respectively, of the following differential equations

$$\frac{p'(x)}{p(x)} = \frac{\alpha a + \beta b - (\alpha + \beta) x}{(b - x)(x - a)}\,,\qquad \frac{p'(x)}{p(x)} = \frac{\alpha + k a - k x}{x - a}\,,\qquad \frac{p'(x)}{p(x)} = h - 2 k x\,.$$

We then see that in the three considered cases, the following two properties hold:

I) $p(x)$ is solution of a differential equation of the type

$$\frac{p'(x)}{p(x)} = \frac{D + E x}{A + B x + C x^2}\,,\qquad (A, B, C, D, E \text{ constants})\ ; \tag{3.3.20}$$

II) the product of $(A + B x + C x^2)\, p(x)$ by any polynomial vanishes for $x = a$ and for $x = b$.

Then we have the following theorem:

Theorem 3.3.V. *In the three mentioned cases the polynomial $P_n(x)$ of the corresponding orthogonal system is solution of the following linear homogeneous differential equation of 2° order*

$$(A + B x + C x^2)\, y'' + [B + D + (2 C + E)\, x]\, y' - n\,[(n + 1)\, C + E]\, y = 0\,. \tag{3.3.21}$$

Proof. Let us consider the following integral

$$I_n = \int_a^b \Pi_{n-1}(x) \frac{d}{dx} [(A + B x + C x^2)\, p(x)\, P_n'(x)]\, dx\,,$$

where $\Pi_{n-1}(x)$ denotes an arbitrary polynomial of degree $n - 1$ and we begin by proving that $I_n = 0$. In fact, operating some integrations by parts and keeping into account properties I) and II) we have

$$\begin{aligned}
I_n &= [\Pi_{n-1}(x)\,(A + B\,x + C\,x^2)\,p(x)\,P_n'(x)]_a^b \\
&\quad - \int_a^b (A + B\,x + C\,x^2)\,p(x)\,P_n'(x)\,\Pi_{n-1}'(x)\,dx \\
&= - \int_a^b (A + B\,x + C\,x^2)\,p(x)\,\Pi_{n-1}'(x)\,P_n'(x)\,dx \\
&= - [(A + B\,x + C\,x^2)\,p(x)\,\Pi_{n-1}'(x)\,P_n(x)]_a^b \\
&\quad + \int_a^b P_n(x)\,[(B + 2\,C\,x)\,p(x)\,\Pi_{n-1}'(x) \\
&\quad + (A + B\,x + C\,x^2)\,p'(x)\,\Pi_{n-1}'(x) \\
&\quad + (A + B\,x + C\,x^2)\,p(x)\,\Pi_{n-1}''(x)]\,dx \\
&= \int_a^b p(x)\,\{[B + D + (2\,C + E)\,x]\,\Pi_{n-1}'(x) \\
&\quad + (A + B\,x + C\,x^2)\,\Pi_{n-1}''(x)\}\,P_n(x)\,dx\,.
\end{aligned}$$

The expression between brackets is a polynomial of degree $\leqq n - 1$ and therefore for (3.3.3) we have $I_n = 0$.

On the other hand, taking the expression of I_n and developing the derivative there occurring we may write

$$\begin{aligned}
I_n = \int_a^b \Pi_{n-1}(x)\,[(B + 2\,C\,x)\,p(x)\,P_n'(x) + (A + B\,x + C\,x^2)\,p'(x)\,P_n'(x) \\
+ (A + B\,x + C\,x^2)\,p(x)\,P_n''(x)]\,dx
\end{aligned}$$

and therefore, taking into account the property I)

$$\begin{aligned}
I_n = \int_a^b p(x)\,\Pi_{n-1}(x)\,\{(A + B\,x + C\,x^2)\,P_n''(x) \\
+ [B + D + (2\,C + E)\,x]\,P_n'(x)\}\,dx\,.
\end{aligned}$$

Wrinting that the latter integral is zero and comparing it with (3.3.3) we see that the polynomial of degree n written between brackets has the same property as $P_n(x)$. Since (3.3.3) determine $P_n(x)$ apart from a constant factor, we may write

$$(A + B\,x + C\,x^2)\,P_n''(x) + [B + D + (2\,C + E)\,x]\,P_n'(x) = c_n\,P_n(x)\,. \quad (3.3.22)$$

Putting $P_n(x) = a_n\,x^n + \cdots$ and equaliing the coefficients of x^n in both members, we find $Cn\,(n - 1)\,a_n + (2\,C + E)\,n\,a_n = c_n\,a_n$ from which $c_n = n\,[(n + 1)\,C + E]$. Therefore (3.3.22) expresses precisely that $P_n(x)$ satisfies (3.3.21) q.e.d.

In § 3.4, 3.6, 3.7 we shall deal with the following particular cases, to which we may apply the preceding theorem:

$$a = -1, \quad b = 1, \quad p(x) = (1-x)^\alpha (1+x)^\beta, \quad (\alpha > -1, \beta > -1),$$
$$\text{(Jacobi polynomials)}, \quad (3.3.23)$$

$$a = 0, \quad b = +\infty, \quad p(x) = x^\alpha e^{-x}, \quad (\alpha > -1),$$
$$\text{(Laguerre polynomials)}, \quad (3.3.24)$$

$$a = -\infty, \quad b = +\infty, \quad p(x) = e^{-x^2}, \quad \text{(Hermite polynomials)}. \quad (3.3.25)$$

3.4. Jacobi polynomials. Jacobi polynomials $P_n^{(\alpha,\beta)}(x)$ are determined (apart from constant factors) by the property of being, in the interval $[-1, 1]$, orthogonal with respect to the weight $(1-x)^\alpha (1+x)^\beta$ [see (3.3.23)].

However they are generally determined by means of the following *Rodrigues formula* (which also fix the factors of proportionality):

$$P_n^{(\alpha,\beta)}(x) = \frac{(-1)^n}{2^n n!} (1-x)^{-\alpha} (1+x)^{-\beta} \frac{d^n}{dx^n} [(1-x)^{\alpha+n} (1+x)^{\beta+n}],$$
$$(n = 0, 1, 2, \ldots); \quad (3.4.1)$$

we shall soon see that property of orthogonality derives from it. Now we observe that from (3.4.1) developping the derivative by the Leibnitz formula and simplyfying, we obtain

$$P_n^{(\alpha,\beta)}(x) = \frac{1}{2^n} \sum_{k=0}^{n} (-1)^{n-k} \binom{\alpha+n}{k} \binom{\beta+n}{n-k} (1-x)^{n-k} (1+x)^k \quad (3.4.2)$$

from which we easily derive the following consequences

$$P_n^{(\alpha,\beta)}(-x) = (-1)^n P_n^{(\beta,\alpha)}(x), \quad (3.4.3)$$

$$P_n^{(\alpha,\beta)}(1) = \binom{\alpha+n}{n}, \quad (3.4.4)$$

$$P_n^{(\alpha,\beta)}(-1) = (-1)^n \binom{\beta+n}{n}. \quad (3.4.5)$$

Putting

$$P_n^{(\alpha,\beta)}(x) = a_n x^n + b_n x^{n-1} + \cdots, \quad (3.4.6)$$

we wish to compute the coefficients a_n, b_n (which sometimes will be denoted by $a_n^{(\alpha,\beta)}$, $b_n^{(\alpha,\beta)}$). We then observe that (3.4.1) can also be written

$$P_n^{(\alpha,\beta)}(x) = \frac{1}{2^n n!} (x-1)^{-\alpha} (x+1)^{-\beta} \frac{d^n}{dx^n} [(x-1)^{\alpha+n} (x+1)^{\beta+n}] \quad (3.4.7)$$

and that we have

$$\begin{aligned}(x-1)^{\alpha+n} (x+1)^{\beta+n} &= x^{\alpha+\beta+2n} \left(1 - \frac{1}{x}\right)^{\alpha+n} \left(1 + \frac{1}{x}\right)^{\beta+n} \\ &= x^{\alpha+\beta+2n} \left(1 - \frac{\alpha+n}{x} + \cdots\right)\left(1 + \frac{\beta+n}{x} + \cdots\right) \\ &= x^{\alpha+\beta+2n} + (\beta-\alpha) x^{\alpha+\beta+2n-1} + \cdots\end{aligned}$$

and therefore, applying (3.2.5)

$$\frac{d^n}{dx^n}[(x-1)^{\alpha+n}(x+1)^{\beta+n}] = \frac{\Gamma(\alpha+\beta+2n+1)}{\Gamma(\alpha+\beta+n+1)} x^{\alpha+\beta+n} + (\beta-\alpha)\frac{\Gamma(\alpha+\beta+2n)}{\Gamma(\alpha+\beta+n)} x^{\alpha+\beta+n-1} + \cdots;$$

we have analogously

$$(x-1)^{-\alpha}(x+1)^{-\beta} = x^{-\alpha-\beta} - (\beta-\alpha)\, x^{-\alpha-\beta-1} + \cdots$$

and therefore (3.4.7) becomes

$$P_n^{(\alpha,\beta)}(x) = \frac{1}{2^n\, n!}[x^{-\alpha-\beta} - (\beta-\alpha)\, x^{-\alpha-\beta-1} + \cdots]\left[\frac{\Gamma(\alpha+\beta+2n+1)}{\Gamma(\alpha+\beta+n+1)} x^{\alpha+\beta+n} + (\beta-\alpha)\frac{\Gamma(\alpha+\beta+2n)}{\Gamma(\alpha+\beta+n)} x^{\alpha+\beta+n-1} + \cdots\right].$$

Developping and using (3.2.4) we are led to

$$P_n^{(\alpha,\beta)}(x) = \frac{1}{2^n\, n!}\left[\frac{\Gamma(\alpha+\beta+2n+1)}{\Gamma(\alpha+\beta+n+1)} x^n - (\beta-\alpha)\, n\, \frac{\Gamma(\alpha+\beta+2n)}{\Gamma(\alpha+\beta+n+1)} x^{n-1} + \cdots\right]$$

whence, comparing it with (3.4.6), we conclude that

$$a_n = \frac{1}{2^n\, n!}\,\frac{\Gamma(\alpha+\beta+2n+1)}{\Gamma(\alpha+\beta+n+1)}, \quad (n = 0, 1, 2, \ldots), \tag{3.4.8}$$

$$b_n = \frac{\alpha-\beta}{2^n\,(n-1)!}\,\frac{\Gamma(\alpha+\beta+2n)}{\Gamma(\alpha+\beta+n+1)}, \quad (n = 1, 2, \ldots). \tag{3.4.9}$$

Now we prove the property of orthogonality of Jacobi polynomials, that is that there exists:

$$\int_{-1}^{1} (1-x)^\alpha (1+x)^\beta P_m^{(\alpha,\beta)}(x)\, P_n^{(\alpha,\beta)}(x)\, dx = 0, \quad (m < n). \tag{3.4.10}$$

It suffices to express $P_n^{(\alpha,\beta)}(x)$ by means of (3.4.1) and operating m integrations by parts, considering that for $k < n$ we obviously have

$$\frac{d^k}{dx^k}[(1-x)^{\alpha+n}(1+x)^{\beta+n}] = 0 \quad \text{for} \quad x = -1 \quad \text{and for} \quad x = 1.$$

In fact the first operation transforms the integral (3.4.10) into the other one

$$\frac{(-1)^n}{2^n\, n!}\int_{-1}^{1} P_m^{(\alpha,\beta)}(x)\,\frac{d^n}{dx^n}[(1-x)^{\alpha+n}(1+x)^{\beta+n}]\, dx$$

which, after the m integrations by parts, becomes

$$\frac{(-1)^{n+m}}{2^n\, n!}\int_{-1}^{1} \frac{d^m}{dx^m} P_m^{(\alpha,\beta)}(x)\,\frac{d^{n-m}}{dx^{n-m}}[(1-x)^{\alpha+n}(1+x)^{\beta+n}]\, dx, \tag{3.4.11}$$

that is using (3.4.6)

$$\frac{(-1)^{n+m}}{2^n\, n!}\, m!\, a_m \left[\frac{d^{n-m-1}}{dx^{n-m-1}}\left[(1-x)^{\alpha+n}(1+x)^{\beta+n}\right]\right]_{-1}^{1} = 0\,.$$

It is thus proved (3.4.10) and it is to be noted that the procedure used to compute the integral (3.4.10) remains valid *up to (3.4.11)* also in the case $m = n$; we namely find

$$\int_{-1}^{1} (1-x)^\alpha (1+x)^\beta \left[P_n^{(\alpha,\beta)}(x)\right]^2 dx$$

$$= \frac{1}{2^n\, n!} \int_{-1}^{1} \frac{d^n}{dx^n} P_n^{(\alpha,\beta)}(x) \cdot (1-x)^{\alpha+n} (1+x)^{\beta+n}\, dx$$

$$= \frac{a_n}{2^n} \int_{-1}^{1} (1-x)^{\alpha+n} (1+x)^{\beta+n}\, dx\,.$$

The latter integral can be computet by substituting $x = 1 - 2t$ which transforms it into $2^{\alpha+\beta+2n+1} \int_0^1 t^{\alpha+n}(1-t)^{\beta+n}\, dt = 2^{\alpha+\beta+2n+1} B(\alpha+n+1, \beta+n+1)$ [see (3.2.9)].

Using (3.2.10) and (3.4.8) we conclude that

$$\int_{-1}^{1} (1-x)^\alpha (1+x)^\beta \left[P_n^{(\alpha,\beta)}(x)\right]^2 dx = h_n^{(\alpha,\beta)} \tag{3.4.12}$$

with

$$h_n^{(\alpha,\beta)} = \frac{1}{2^n} \cdot \frac{1}{2^n\, n!}\, \frac{\Gamma(\alpha+\beta+2n+1)}{\Gamma(\alpha+\beta+n+1)} \cdot 2^{\alpha+\beta+2n+1}\, \frac{\Gamma(\alpha+n+1)\,\Gamma(\beta+n+1)}{\Gamma(\alpha+\beta+2n+2)}$$

that is, simplyfying and using (3.2.4)

$$h_n^{(\alpha,\beta)} = \frac{2^{\alpha+\beta+1}}{\alpha+\beta+2n+1}\, \frac{\Gamma(\alpha+n+1)\,\Gamma(\beta+n+1)}{n!\,\Gamma(\alpha+\beta+n+1)}\,. \tag{3.4.13}$$

Equation (3.4.12) shows that the orthogonal system of the Jacobi polynomials *is not normal.* In order to have an orthonormal system we evidently need to divide $P_n^{(\alpha,\beta)}(x)$ by the constant $\sqrt{h_n^{(\alpha,\beta)}}$; we may therefore state that the *system*

$$\frac{P_n^{(\alpha,\beta)}(x)}{\sqrt{h_n^{(\alpha,\beta)}}}, \quad (n = 0, 1, 2, \ldots) \tag{3.4.14}$$

is orthonormal in $[-1, 1]$ *with respect to the weight* $(1-x)^\alpha(1+x)^\beta$. We may now derive for the Jacobi polynomials the *Christoffel-Darboux formula,* applying to the system (3.4.14) the formula (3.3.15). We find

$$\sum_{k=0}^{n} \frac{P_k^{(\alpha,\beta)}(x)}{\sqrt{h_k^{(\alpha,\beta)}}}\, \frac{P_k^{(\alpha,\beta)}(y)}{\sqrt{h_k^{(\alpha,\beta)}}} = \frac{a_n/\sqrt{h_n^{(\alpha,\beta)}}}{a_{n+1}/\sqrt{h_{n+1}^{(\alpha,\beta)}}}\, \frac{\dfrac{P_{n+1}^{(\alpha,\beta)}(x)}{\sqrt{h_{n+1}^{(\alpha,\beta)}}}\, \dfrac{P_n^{(\alpha,\beta)}(y)}{\sqrt{h_n^{(\alpha,\beta)}}} - \dfrac{P_{n+1}^{(\alpha,\beta)}(y)}{\sqrt{h_{n+1}^{(\alpha,\beta)}}}\, \dfrac{P_n^{(\alpha,\beta)}(x)}{\sqrt{h_n^{(\alpha,\beta)}}}}{x-y}$$

that is, simplyfying and using (3.4.8)

$$\sum_{k=0}^{n} \frac{P_k^{(\alpha,\beta)}(x)\, P_k^{(\alpha,\beta)}(y)}{h_k^{(\alpha,\beta)}} = \frac{A_n^{(\alpha,\beta)}}{h_n^{(\alpha,\beta)}} \frac{P_{n+1}^{(\alpha,\beta)}(x)\, P_n^{(\alpha,\beta)}(y) - P_{n+1}^{(\alpha,\beta)}(y)\, P_n^{(\alpha,\beta)}(x)}{x-y} \quad (3.4.15)$$

with [see (3.3.11)]

$$A_n^{(\alpha,\beta)} = \frac{a_n}{a_{n+1}} = \frac{2\,(n+1)\,(\alpha+\beta+n+1)}{(\alpha+\beta+2\,n+1)\,(\alpha+\beta+2\,n+2)}\,. \quad (3.4.16)$$

We may also derive the *differential equation satisfied by the Jacobi polynomials*. It suffices to apply theorem 3.3.V considering that (3.3.20) now is written $\frac{p'(x)}{p(x)} = \frac{\beta-\alpha-(\alpha+\beta)\,x}{1-x^2}$ that is that we have $A = 1$, $B = 0$, $C = -1$, $D = \beta - \alpha$, $E = -(\alpha+\beta)$. With that (3.3.21) becomes

$$(1-x^2)\,y'' + [\beta-\alpha-(\alpha+\beta+2)\,x]\,y' + n\,(\alpha+\beta+n+1)\,y = 0\,. \quad (3.4.17)$$

We have only to deduce for the Jacobi polynomials the *recurrence formula* of type (3.3.10). In virtue of (3.3.11) and (3.4.8) we first find [see (3.4.16)]:

$$A_n = \frac{2\,(n+1)\,(\alpha+\beta+n+1)}{(\alpha+\beta+2\,n+1)\,(\alpha+\beta+2\,n+2)}\,;$$

successively from (3.3.14), (3.4.8) and (3.4.9) we easily deduce

$$B_n = -\frac{\alpha^2-\beta^2}{(\alpha+\beta+2\,n)\,(\alpha+\beta+2\,n+2)}\,;$$

finally from (3.3.13), (3.4.8), (3.4.12) and (3.4.13) we obtain with a long computation

$$C_n = \frac{2\,(\alpha+n)\,(\beta+n)}{(\alpha+\beta+2\,n)\,(\alpha+\beta+2\,n+1)}\,.$$

Therefore we write (3.3.10):

$$\begin{aligned} x\,P_n^{(\alpha,\beta)}(x) = {} & \frac{1}{(\alpha+\beta+2\,n)\,(\alpha+\beta+2\,n+1)\,(\alpha+\beta+2\,n+2)} \times \\ & \times [2\,(n+1)\,(\alpha+\beta+n+1)\,(\alpha+\beta+2\,n)\,P_{n+1}^{(\alpha,\beta)}(x) \\ & - (\alpha^2-\beta^2)\,(\alpha+\beta+2\,n+1)\,P_n^{(\alpha,\beta)}(x) \\ & + 2\,(\alpha+n)\,(\beta+n)\,(\alpha+\beta+2\,n+2)\,P_{n-1}^{(\alpha,\beta)}(x)] \end{aligned}$$

but it is convenient to write it in the form

$$\begin{aligned} & 2\,(n+1)\,(\alpha+\beta+n+1)\,(\alpha+\beta+2\,n)\,P_{n+1}^{(\alpha,\beta)}(x) \\ & \quad = (\alpha+\beta+2\,n+1)\,[(\alpha+\beta+2\,n)\,(\alpha+\beta+2\,n+2)\,x \\ & \qquad + \alpha^2-\beta^2]\,P_n^{(\alpha,\beta)}(x) - 2\,(\alpha+n)\,(\beta+n)\,(\alpha+\beta+2\,n+2)\,P_{n-1}^{(\alpha,\beta)}(x)\,, \\ & \qquad (n = 1, 2, 3, \ldots)\,. \end{aligned} \quad (3.4.18)$$

Formula (3.4.18) is of use for computing for recurrence all the Jacobi polynomials starting form $P_0^{(\alpha,\beta)}(x)$, $P_1^{(\alpha,\beta)}(x)$ for which (3.4.1) immediately yields:

$$P_0^{(\alpha,\beta)}(x) = 1\,, \quad P_1^{(\alpha,\beta)}(x) = \frac{1}{2}\,[(\alpha+\beta+2)\,x + \alpha - \beta]\,. \quad (3.4.19)$$

We shall add some other formulae on Jacobi polynomials that will be useful in the following:

Differentiating (3.4.17) and putting $y' = z$ we obtain the new differential equation

$$(1 - x^2)\, z'' + [\beta - \alpha - (\alpha + \beta + 4)\, x]\, z' + (n - 1)(\alpha + \beta + n + 2)\, z = 0\,, \tag{3.4.20}$$

which for its structure admits of the polynomial solution $\frac{d}{dx} P_n^{(\alpha,\beta)}(x)$. But (3.4.20) can also be considered as derived from (3.4.17) changing y into z, n into $n - 1$, α into $\alpha + 1$, β into $\beta + 1$ and therefore it admits also of the polynomial solution $P_{n-1}^{(\alpha+1,\beta+1)}(x)$. There follows that $\frac{d}{dx} P_n^{(\alpha,\beta)}(x) = c_n P_{n-1}^{(\alpha+1,\beta+1)}(x)$ where c_n denotes a constant whose value can be determined by equalizing the coefficients of x^{n-1} in both members. We find $n\, a_n^{(\alpha,\beta)} = c_n\, a_{n-1}^{(\alpha+1,\beta+1)}$ wherefrom, using (3.4.8), we obtain $c_n = \frac{1}{2}(\alpha + \beta + n + 1)$. Therefore the following formula holds

$$\frac{d}{dx} P_n^{(\alpha,\beta)}(x) = \frac{1}{2}(\alpha + \beta + n + 1)\, P_{n-1}^{(\alpha+1,\beta+1)}(x)\,. \tag{3.4.21}$$

Now we prove these two other formulae

$$(\alpha + \beta + 2n)\, P_n^{(\alpha-1,\beta)}(x) = (\alpha + \beta + n)\, P_n^{(\alpha,\beta)}(x) - (\beta + n)\, P_{n-1}^{(\alpha,\beta)}(x)\,, \tag{3.4.22}$$

$$(\alpha + \beta + 2n)\, P_n^{(\alpha,\beta-1)}(x) = (\alpha + \beta + n)\, P_n^{(\alpha,\beta)}(x) + (\alpha + n)\, P_{n-1}^{(\alpha,\beta)}(x)\,, \tag{3.4.23}$$

from which by subtracting there follows

$$P_n^{(\alpha,\beta-1)}(x) - P_n^{(\alpha-1,\beta)}(x) = P_{n-1}^{(\alpha,\beta)}(x)\,. \tag{3.4.24}$$

It suffices to prove (3.4.22) since (3.4.23) can be deduced from it changing α by β, x by $-x$ and then applying (3.4.3). To prove (3.4.22) we note that we evidently may write the polynomial $P_n^{(\alpha-1,\beta)}(x)$ in the following way

$$P_n^{(\alpha-1,\beta)}(x) = A\, P_n^{(\alpha,\beta)}(x) + B\, P_{n-1}^{(\alpha,\beta)}(x) + \sum_{k=2}^{n} C_k\, P_{n-k}^{(\alpha,\beta)}(x)$$

where A, B, C_k are constants. Then multiplying both members by $(1 - x)^\alpha (1 + x)^\beta P_m^{(\alpha,\beta)}(x)$, $(m = 0, 1, \ldots, n - 2)$ and then integrating on $[-1, 1]$ we soon see that, for the orthogonality property, all the coefficients C_k are zero. Therefore we have

$$P_n^{(\alpha-1,\beta)}(x) = A\, P_n^{(\alpha,\beta)}(x) + B\, P_{n-1}^{(\alpha,\beta)}(x) \tag{3.4.25}$$

and to determine A, B it suffices to put $x = 1$ and $x = -1$. Using (3.4.4) and (3.4.5) we find

$$\binom{\alpha + n}{n} A + \binom{\alpha + n - 1}{n - 1} B = \binom{\alpha + n - 1}{n},$$

$$(-1)^n \binom{\beta + n}{n} A + (-1)^{n-1} \binom{\beta + n - 1}{n - 1} B = (-1)^n \binom{\beta + n}{n}$$

that is

$$(\alpha + n) A + n B = \alpha , \quad (\beta + n) A - n B = \beta + n ;$$

we obtain $A = \dfrac{\alpha + \beta + n}{\alpha + \beta + 2n}$, $B = -\dfrac{\beta + n}{\alpha + \beta + 2n}$ and substituting in (3.4.25) we have (3.4.22).

Analogously we may prove

$$\frac{1}{2}(1 - x)(\alpha + \beta + 2n + 2) P_n^{(\alpha+1, \beta)}(x) \\ = -(n + 1) P_{n+1}^{(\alpha, \beta)}(x) + (\alpha + n + 1) P_n^{(\alpha, \beta)}(x) , \qquad (3.4.26)$$

$$\frac{1}{2}(1 + x)(\alpha + \beta + 2n + 2) P_n^{(\alpha, \beta+1)}(x) \\ = (n + 1) P_{n+1}^{(\alpha, \beta)}(x) + (\beta + n + 1) P_n^{(\alpha, \beta)}(x) , \qquad (3.4.27)$$

from which by summing there follows

$$(1 - x) P_n^{(\alpha+1, \beta)}(x) + (1 + x) P_n^{(\alpha, \beta+1)}(x) = 2 P_n^{(\alpha, \beta)}(x) ; \qquad (3.4.28)$$

we leave to the reader the easy computations.

Now we prove these other formulae:

$$(\alpha + \beta + 2n)(1 - x^2) \frac{d}{dx} P_n^{(\alpha, \beta)}(x) \\ = -n [(\alpha + \beta + 2n) x + \beta - \alpha] P_n^{(\alpha, \beta)}(x) \\ + 2(\alpha + n)(\beta + n) P_{n-1}^{(\alpha, \beta)}(x) , \qquad (3.4.29)$$

$$(\alpha + \beta + 2n + 2)(1 - x^2) \frac{d}{dx} P_n^{(\alpha, \beta)}(x) \\ = (\alpha + \beta + n + 1) [(\alpha + \beta + 2n + 2) x + \alpha - \beta] P_n^{(\alpha, \beta)}(x) \\ - 2(n + 1)(\alpha + \beta + n + 1) P_{n+1}^{(\alpha, \beta)}(x) . \qquad (3.4.30)$$

It suffices to prove the former since the latter derives from it substituting $P_{n-1}^{(\alpha, \beta)}(x)$ by the expressions that can be obtained from the recurrence formula (3.4.18).

To prove (3.4.29) we note that $(1 - x^2) \dfrac{d}{dx} P_n^{(\alpha, \beta)}(x) + n x P_n^{(\alpha, \beta)}(x)$ is a polynomial of degree n and that as such it may be written

$$(1 - x^2) \frac{d}{dx} P_n^{(\alpha, \beta)}(x) + n x P_n^{(\alpha, \beta)}(x) \\ = A P_n^{(\alpha, \beta)}(x) + B P_{n-1}^{(\alpha, \beta)}(x) + \sum_{k=2}^{n} C_k P_{n-k}^{(\alpha, \beta)}(x) \qquad (3.4.31)$$

where A, B, C_k are constants. Multiplying both members by $(1 - x)^\alpha \times (1 + x)^\beta P_m^{(\alpha, \beta)}(x)$, $(m = 0, 1, \ldots, n - 2)$ and then integrating on $[-1, 1]$ we deduce, keeping also into account (3.4.21), that all the coefficients C_k are zero, so that (3.4.31) becomes

$$(1 - x^2) \frac{d}{dx} P_n^{(\alpha, \beta)}(x) + n x P_n^{(\alpha, \beta)}(x) = A P_n^{(\alpha, \beta)} + B P_{n-1}^{(\alpha, \beta)}(x) . \qquad (3.4.32)$$

To determine A and B it suffices to put $x = 1$ and $x = -1$ and keep into account (3.4.4), (3.4.5). We find

$$\binom{\alpha+n}{n} A + \binom{\alpha+n-1}{n-1} B = n \binom{\alpha+n}{n},$$

$$(-1)^n \binom{\beta+n}{n} A + (-1)^{n-1} \binom{\beta+n-1}{n-1} B = -n(-1)^n \binom{\beta+n}{n}$$

that is

$$(\alpha+n) A + n B = n(\alpha+n), \quad (\beta+n) A - n B = -n(\beta+n).$$

We obtain $A = \dfrac{n(\alpha-\beta)}{\alpha+\beta+2n}$, $B = \dfrac{2(\alpha+n)(\beta+n)}{\alpha+\beta+2n}$ and substituting in (3.4.32) we easily obtain (3.4.29).

We shall still need the properties expressed by

$$\sum_{k=0}^{n} \frac{(\alpha+\beta+2k+1)\,\Gamma(\alpha+\beta+k+1)}{\Gamma(\beta+k+1)} P_k^{(\alpha,\beta)}(x) = \frac{\Gamma(\alpha+\beta+n+2)}{\Gamma(\beta+n+1)} P_n^{(\alpha+1,\beta)}(x), \tag{3.4.33}$$

$$\sum_{k=0}^{n} (-1)^k \frac{(\alpha+\beta+2k+1)\,\Gamma(\alpha+\beta+k+1)}{\Gamma(\alpha+k+1)} P_k^{(\alpha,\beta)}(x)$$

$$= (-1)^n \frac{\Gamma(\alpha+\beta+n+2)}{\Gamma(\alpha+n+1)} P_n^{(\alpha,\beta+1)}(x), \tag{3.4.34}$$

of which it suffices to prove the first one, since from it we pass to the second one changing α by β, x by $-x$ and applying (3.4.3).

To prove (3.4.33) we put $y = 1$ in the Christoffel-Darboux formula (3.4.15). Using (3.4.4) and (3.4.13) the first member easily becomes

$$\frac{1}{2^{\alpha+\beta+1}\,\Gamma(\alpha+1)} \sum_{k=0}^{n} \frac{(\alpha+\beta+2k+1)\,\Gamma(\alpha+\beta+k+1)}{\Gamma(\beta+k+1)} P_k^{(\alpha,\beta)}(x), \tag{3.4.35}$$

while, using (3.4.4), (3.4.13) and (3.4.16) the second member becomes

$$\frac{1}{2^{\alpha+\beta+1}\,\Gamma(\alpha+1)} \cdot \frac{\Gamma(\alpha+\beta+n+2)}{\Gamma(\beta+n+1)} \cdot \frac{2}{\alpha+\beta+2n+2} \times$$

$$\times \frac{(n+1)\,P_{n+1}^{(\alpha,\beta)}(x) - (\alpha+n+1)\,P_n^{(\alpha,\beta)}(x)}{x-1}$$

that is, expressing the last numerator by means of (3.4.26):

$$\frac{1}{2^{\alpha+\beta+1}\,\Gamma(\alpha+1)} \frac{\Gamma(\alpha+\beta+n+2)}{\Gamma(\beta+n+1)} P_n^{(\alpha+1,\beta)}(x). \tag{3.4.36}$$

Equalizing the two expressions (3.4.35) and (3.4.36) we obtain (3.4.33).

3.5. Particular cases of Jacobi polynomials. The most important particular case of the Jacobi polynomials is the one in which we have $\alpha = \beta$. Usually we put

$$\alpha = \beta = \lambda - \frac{1}{2}, \qquad \left(\lambda > -\frac{1}{2}\right), \tag{3.5.1}$$

and, assuming $\lambda \neq 0$, we change the polynomial $P_n^{\left(\lambda-\frac{1}{2},\lambda-\frac{1}{2}\right)}(x)$ by a certain constant factor, so arriving to the so-called *ultraspherical polynomials* or *Gegenbauer polynomials* $P_n^{(\lambda)}(x)$ determined by the formula

$$P_n^{(\lambda)}(x) = \frac{\Gamma\left(\lambda+\frac{1}{2}\right)\Gamma(2\lambda+n)}{\Gamma(2\lambda)\,\Gamma\left(\lambda+n+\frac{1}{2}\right)}\,P_n^{\left(\lambda-\frac{1}{2},\,\lambda-\frac{1}{2}\right)}(x)\,, \qquad \left(\lambda > -\frac{1}{2},\, \lambda \neq 0\right),^{1)} \tag{3.5.2}$$

which, using (3.2.5), can be transformed into

$$P_n^{(\lambda)}(x) = \frac{2\lambda(2\lambda+1)\dots(2\lambda+n-1)}{\left(\lambda+\frac{1}{2}\right)\left(\lambda+\frac{3}{2}\right)\dots\left(\lambda+\frac{2n-1}{2}\right)}\,P_n^{\left(\lambda-\frac{1}{2},\,\lambda-\frac{1}{2}\right)}(x)\,. \tag{3.5.3}$$

The definition is usually extended to the case $\lambda = 0$, putting $P_n^{(0)}(x) = \lim\limits_{\lambda\to 0}\frac{P_n^{(\lambda)}(x)}{\lambda}$; the passage to the limit is soon operated using (3.5.3) and so obtaining

$$P_n^{(0)}(x) = \lim_{\lambda\to 0}\frac{P_n^{(\lambda)}(x)}{\lambda} = \frac{2\,(n-1)!}{\frac{(2n-1)!!}{2^n}}\,P_n^{\left(-\frac{1}{2},\,-\frac{1}{2}\right)}(x)$$

$$= \frac{2}{n}\,\frac{(2n)!!}{(2n-1)!!}\,P_n^{\left(-\frac{1}{2},\,-\frac{1}{2}\right)}(x)\,. \tag{3.5.4}$$

We leave to the reader the task of deducing from the formulae of the § 3.4 those concerning the ultraspherical polynomials. We shall confine ourselves to observe that to such polynomials we may apply theorem 3.3.IV $\left[\text{since } p(x) = (1-x^2)^{\lambda-\frac{1}{2}}\right]$, whence $P_n^{(\lambda)}(x)$ is *an even or odd function according to that n is even or odd*; there follows that *the zeros of $P_n^{(\lambda)}(x)$ are set simmetrically with respect to the origin.*

Now we examine some particular cases of ultraspherical polynomials. The most known is the one of *spherical polynomials* or *Legendre polynomials* $P_n(x)$ which can be obtained from (3.5.2) putting $\lambda = \frac{1}{2}$ (that is $\alpha = \beta = 0$); therefore we have, also recalling (3.4.1):

$$P_n(x) = P_n^{\left(\frac{1}{2}\right)}(x) = P_n^{(0,\,0)}(x) = \frac{(-1)^n}{2^n\,n!}\,\frac{d^n}{dx^n}(1-x^2)^n\,. \tag{3.5.5}$$

[1]) In § 3.2 we have defined the function $\Gamma(\alpha)$ only for $\alpha > 0$; since here occurs also the factor $\Gamma(2\lambda)$ we are forced to consider this function for $-1 < \alpha < 0$. The extension of $\Gamma(\alpha)$ to *non-integer negative* values of α can be obtained in the way shown in § 3.10, Problem 7.

Legendre polynomials are orthogonal with respect to the weight function 1; then we may write

$$\int_{-1}^{1} P_m(x)\, P_n(x)\, dx = \begin{cases} 0 & (m \neq n)\,, \\ \dfrac{2}{2\,n+1} & (m = n)\,, \end{cases} \tag{3.5.6}$$

having also used (3.4.12) and (3.4.13).

Another case particularly interesting is the one in which $\lambda = 1$; then we have the so-called *Tchebychef polynomials of second kind* $V_n(x)$ determined by

$$V_n(x) = P_n^{(1)}(x) = \frac{2 \cdot 3 \ldots (n+1)}{\frac{3}{2} \cdot \frac{5}{2} \cdots \frac{2\,n+1}{2}}\, P_n^{\left(\frac{1}{2}, \frac{1}{2}\right)}(x) = \frac{1}{2}\, \frac{(2\,n+2)!!}{(2\,n+1)!!}\, P_n^{\left(\frac{1}{2}, \frac{1}{2}\right)}(x)\,, \tag{3.5.7}$$

as it follows from (3.5.3). But for $V_n(x)$ we may give another very simple expression; we wish to prove that, putting $x = \cos\theta$, we have

$$V_n(\cos\theta) = \frac{\sin(n+1)\,\theta}{\sin\theta}\,. \tag{3.5.8}$$

In fact, considering again (3.4.18) and (3.4.19) and putting in it $\alpha = \beta = \frac{1}{2}$, we may say that the polynomials $P_n^{\left(\frac{1}{2}, \frac{1}{2}\right)}(x)$ are determined by

$$P_0^{\left(\frac{1}{2}, \frac{1}{2}\right)}(x) = 1\,, \qquad P_1^{\left(\frac{1}{2}, \frac{1}{2}\right)}(x) = \frac{3}{2}\, x\,,$$

$$2\,(n+1)\,(n+2)\, P_{n+1}^{\left(\frac{1}{2}, \frac{1}{2}\right)}(x) = (2\,n+2)\,(2\,n+3)\, x\, P_n^{\left(\frac{1}{2}, \frac{1}{2}\right)}(x) - \frac{1}{2}\,(2\,n+1)\,(2\,n+3)\, P_{n-1}^{\left(\frac{1}{2}, \frac{1}{2}\right)}(x)\,.$$

If we transform these equations by expressing $P_n^{\left(\frac{1}{2}, \frac{1}{2}\right)}(x)$ by means of $V_n(x)$ using (3.5.7), we see that the polynomials $V_n(x)$ can be considered fully determined by

$$\left.\begin{aligned} V_0(x) &= 1\,, \qquad V_1(x) = 2\,x\,, \\ V_{n+1}(x) &= 2\,x\, V_n(x) - V_{n-1}(x)\,. \end{aligned}\right\} \tag{3.5.9}$$

Having put $x = \cos\theta$ we immediately see that (3.5.9) are satisfied putting in the place of $V_n(\cos\theta)$ the expression given by (3.5.8), q.e.d.

We therefore have the formulae

$$\left.\begin{aligned} V_n(x) &= \frac{\sin[(n+1)\arccos x]}{\sqrt{1-x^2}}\,, \\ P_n^{\left(\frac{1}{2}, \frac{1}{2}\right)}(x) &= 2\, \frac{(2\,n+1)!!}{(2\,n+2)!!}\, \frac{\sin(n+1)\,\theta}{\sin\theta} \quad (\text{with } x = \cos\theta)\,. \end{aligned}\right\} \tag{3.5.10}$$

Also for the polynomials $P_n^{(0)}(x)$ determined by (3.5.4) it is possible to give another expression which joins them to the *Tchebychef polynomials of first kind* $T_n(x) = \cos n\,\theta$. Namely we want to prove that we have

$$P_n^{(0)}(x) = \frac{2}{n}\,T_n(x) = \frac{2}{n}\cos n\,\theta = \frac{2}{n}\cos(n \arccos x)\,. \qquad (3.5.11)$$

In fact, operating as in the preceding case, we see that the polynomials $P_n^{\left(-\frac{1}{2},-\frac{1}{2}\right)}(x)$ are determined by

$$P_0^{\left(-\frac{1}{2},-\frac{1}{2}\right)}(x) = 1\,, \qquad P_1^{\left(-\frac{1}{2},-\frac{1}{2}\right)}(x) = \frac{1}{2}\,x$$

$$2\,n\,(n+1)\,P_{n+1}^{\left(-\frac{1}{2},-\frac{1}{2}\right)}(x) = 2\,n\,(2\,n+1)\,x\,P_n^{\left(-\frac{1}{2},-\frac{1}{2}\right)}(x)$$
$$-\frac{1}{2}(2\,n-1)(2\,n+1)\,P_{n-1}^{\left(-\frac{1}{2},-\frac{1}{2}\right)}(x)$$

and therefore expressing $P_n^{\left(-\frac{1}{2},-\frac{1}{2}\right)}(x)$ by means of $T_n(x)$ using (3.5.4) and (3.5.11), that the polynomials $T_n(x)$ are fully determined by

$$T_0(x) = 1\,, \qquad T_1(x) = x\,,$$
$$T_{n+1}(x) = 2.x\,T_n(x) - T_{n-1}(x)\,.$$

But, with $x = \cos\theta$, we soon see that these formulae are satisfied by $T_n(x) = \cos n\,\theta$ q.e.d.

Other particularly interesting Jacobi polynomials are those with $\alpha = \frac{1}{2}$, $\beta = -\frac{1}{2}$ or $\alpha = -\frac{1}{2}$, $\beta = \frac{1}{2}$. We prove that we have

$$P_n^{\left(\frac{1}{2},-\frac{1}{2}\right)}(x) = \frac{(2\,n-1)!!}{(2\,n)!!}\,\frac{\sin\left[\left(n+\frac{1}{2}\right)\arccos x\right]}{\sin\left(\frac{1}{2}\arccos x\right)}\,, \qquad (3.5.12)$$

$$P_n^{\left(-\frac{1}{2},\frac{1}{2}\right)}(x) = \frac{(2\,n-1)!!}{(2\,n)!!}\,\frac{\cos\left[\left(n+\frac{1}{2}\right)\arccos x\right]}{\cos\left(\frac{1}{2}\arccos x\right)}\,. \qquad (3.5.13)$$

It suffices to prove the first equation since (3.5.13) is soon deduced from (3.5.12) changing x by $-x$ (and therefore $\arccos x$ by π-$\arccos x$) and applying (3.4.3). Let us consider again (3.4.23) and write it with $\alpha = \beta = \frac{1}{2}$; we thus obtain

$$(2\,n+1)\,P_n^{\left(\frac{1}{2},-\frac{1}{2}\right)}(x) = (n+1)\,P_n^{\left(\frac{1}{2},\frac{1}{2}\right)}(x) + \left(n+\frac{1}{2}\right)P_{n-1}^{\left(\frac{1}{2},\frac{1}{2}\right)}(x)$$

and therefore, using the second equation of (3.5.10)

$$(2n+1)\,P_n^{\left(\frac{1}{2},\,-\frac{1}{2}\right)}(x) = (n+1)\,2\,\frac{(2n+1)!!}{(2n+2)!!}\,\frac{\sin(n+1)\theta}{\sin\theta} + \frac{2n+1}{2}\cdot 2\,\frac{(2n-1)!!}{(2n)!!}\,\frac{\sin n\theta}{\sin\theta}.$$

There follows

$$P_n^{\left(\frac{1}{2},\,-\frac{1}{2}\right)}(x) = \frac{(2n-1)!!}{(2n)!!}\,\frac{\sin(n+1)\theta+\sin n\theta}{\sin\theta} = \frac{(2n-1)!!}{(2n)!!}\,\frac{2\sin\left(n+\frac{1}{2}\right)\theta\cos\frac{\theta}{2}}{2\sin\frac{\theta}{2}\cos\frac{\theta}{2}}$$

wherefrom (3.5.12).

3.6. Laguerre polynomials. Laguerre polynomials $L_n^{(\alpha)}(x)$ (with $\alpha > -1$) are determined (apart from constant factors) by the property of being, on the interval $[0, +\infty)$, orthogonal with respect to the weight $x^\alpha e^{-x}$ [see (3.3.24)]. However it is usually determined directly by means of the *Rodrigues formula*:

$$L_n^{(\alpha)}(x) = \frac{1}{n!}\,x^{-\alpha}\,e^x\,\frac{d^n}{dx^n}(x^{\alpha+n}\,e^{-x}), \qquad (n = 0, 1, 2, \ldots), \tag{3.6.1}$$

from which we easily deduce

$$L_n^{(\alpha)}(x) = \sum_{k=0}^{n}(-1)^k\binom{\alpha+n}{n-k}\frac{x^k}{k!}. \tag{3.6.2}$$

Therefore, putting

$$L_n^{(\alpha)}(x) = a_n\,x^n + b_n\,x^{n-1} + \cdots, \tag{3.6.3}$$

we have

$$a_n = \frac{(-1)^n}{n!}, \tag{3.6.4}$$

$$b_n = (-1)^{n-1}\frac{\alpha+n}{(n-1)!}. \tag{3.6.5}$$

We now prove the property of orthogonality of Laguerre polynomials, expressed by

$$\int_0^{+\infty} x^\alpha\,e^{-x}\,L_m^{(\alpha)}(x)\,L_n^{(\alpha)}(x)\,dx = 0, \qquad (m < n). \tag{3.6.6}$$

It is sufficient to express $L_n^{(\alpha)}(x)$ by means of (3.6.1) and then operate m integrations by parts to transform integral (3.6.6) into

$$\frac{(-1)^m}{n!}\int_0^{+\infty}\frac{d^m}{dx^m}L_m^{(\alpha)}(x)\,\frac{d^{n-m}}{dx^{n-m}}(x^{\alpha+n}\,e^{-x})\,dx = \frac{(-1)^m}{n!}\,m!\,a_m\left[\frac{d^{n-m-1}}{dx^{n-m-1}}(x^{\alpha+n}\,e^{-x})\right]_0^{+\infty} = 0.$$

If on the contrary $m = n$ the same procedure yields

$$\int_0^{+\infty} x^\alpha e^{-x} [L_n^{(\alpha)}(x)]^2 dx = \frac{(-1)^n}{n!} n!\, a_n \int_0^{+\infty} x^{\alpha+n} e^{-x} dx ,$$

but for (3.6.4) and (3.2.1) this last expression has value $(-1)^n \frac{(-1)^n}{n!} \Gamma(\alpha+n+1)$ whence we conclude

$$\int_0^{+\infty} x^\alpha e^{-x} [L_n^{(\alpha)}(x)]^2 dx = \frac{\Gamma(\alpha+n+1)}{n!} \tag{3.6.7}$$

and therefore that the *system*

$$L_n^{(\alpha)}(x) \Big/ \sqrt{\frac{\Gamma(\alpha+n+1)}{n!}} , \qquad (n = 0, 1, 2, \ldots) \tag{3.6.8}$$

is orthonormal in $[0, +\infty)$ *with respect to the weight function* $x^\alpha e^{-x}$. Applying to this system the *Christoffel-Darboux formula* (3.3.15) we easily obtain

$$\sum_{k=0}^{n} \frac{k!}{\Gamma(\alpha+k+1)} L_k^{(\alpha)}(x) L_k^{(\alpha)}(y) = -\frac{(n+1)!}{\Gamma(\alpha+n+1)} \frac{L_{n+1}^{(\alpha)}(x) L_n^{(\alpha)}(y) - L_{n+1}^{(\alpha)}(y) L_n^{(\alpha)}(x)}{x-y} . \tag{3.6.9}$$

Now we deduce the differential equation satisfied by Laguerre polynomials. It suffices to apply theorem 3.3.V considering that now (3.3.20) is written $\frac{p'(x)}{p(x)} = \frac{\alpha - x}{x}$, that is that we have $A = 0$, $B = 1$, $C = 0$, $D = \alpha$, $E = -1$. Thus (3.3.21) becomes

$$x y'' + (\alpha + 1 - x) y' + n y = 0 . \tag{3.6.10}$$

We have only to deduce for Laguerre polynomials the *recurrence formula* of the type (3.3.10), computing the coefficients A_n, B_n, C_n of such a formula. Using (3.3.11) and (3.6.4) we find $A_n = -(n+1)$; from (3.3.14), (3.6.4), (3.6.5) we obtain then $B_n = \alpha + 2n + 1$; finally from (3.3.13), (3.6.4) and (3.6.7) we deduce $C_n = -(\alpha + n)$. Therefore (3.3.10) gives

$$(n+1) L_{n+1}^{(\alpha)}(x) = (\alpha + 2n + 1 - x) L_n^{(\alpha)}(x) - (\alpha + n) L_{n-1}^{(\alpha)}(x) , \quad (n = 1, 2, \ldots) , \tag{3.6.11}$$

which allows the computation of all polynomials $L_n^{(\alpha)}(x)$ starting from the first two for which (3.6.1) yields $L_0^{(\alpha)}(x) = 1$, $L_1^{(\alpha)}(x) = -x + \alpha + 1$. We add at last these other two formulae

$$(n+1) L_{n+1}^{(\alpha)}(x) = x \frac{d}{dx} L_n^{(\alpha)}(x) + (\alpha + n + 1 - x) L_n^{(\alpha)}(x) , \tag{3.6.12}$$

$$x \frac{d}{dx} L_n^{(\alpha)}(x) = n L_n^{(\alpha)}(x) - (\alpha + n) L_{n-1}^{(\alpha)}(x) . \tag{3.6.13}$$

We need to prove only the first equation, since the second equation can be obtained by subtracting member by member (3.6.11) and (3.6.12). If we use (3.6.1) we may write

$$(n+1)\,L_{n+1}^{(\alpha)}(x) = (n+1)\frac{1}{(n+1)!}\,x^{-\alpha}\,e^{x}\,\frac{d^{n+1}}{dx^{n+1}}\,(x^{\alpha+n+1}\,e^{-x})$$

$$= \frac{1}{n!}\,x^{-\alpha}\,e^{x}\left[x\,\frac{d^{n+1}}{dx^{n+1}}\,(x^{\alpha+n}\,e^{-x}) + (n+1)\frac{d^{n}}{dx^{n}}\,(x^{\alpha+n}\,e^{-x})\right]$$

$$= \frac{1}{n!}\,x^{-\alpha}\,e^{x}\left\{x\,\frac{d}{dx}\,[n!\,x^{\alpha}\,e^{-x}\,L_n^{(\alpha)}(x)] + (n+1)\,n!\,x^{\alpha}\,e^{-x}\,L_n^{(\alpha)}(x)\right\}$$

$$= x^{-\alpha}\,e^{x}\left\{x\left[\alpha\,x^{\alpha-1}\,e^{-x}\,L_n^{(\alpha)}(x) - x^{\alpha}\,e^{-x}\,L_n^{(\alpha)}(x)\right.\right.$$

$$\left.\left.+\,x^{\alpha}\,e^{-x}\,\frac{d}{dx}\,L_n^{(\alpha)}(x)\right] + (n+1)\,x^{\alpha}\,e^{-x}\,L_n^{(\alpha)}(x)\right\}$$

$$= \alpha\,L_n^{(\alpha)}(x) - x\,L_n^{(\alpha)}(x) + x\,\frac{d}{dx}\,L_n^{(\alpha)}(x) + (n+1)\,L_n^{(\alpha)}(x)$$

and this last expression coincides with the second member of (3.6.12).

3.7. Hermite polynomials. The Hermite polynomials $H_n(x)$ are determined (apart from constant factors) by the property of being on the interval $(-\infty, +\infty)$ orthogonal with respect to the weight e^{-x^2} [see (3.3.25)]. We may apply to them theorem 3.3.IV; therefore *$H_n(x)$ is an even or odd function according to if n is even or odd and its zeros are set symmetrically with respect to the origin.* They can be defined starting from the *Rodrigues formula*

$$H_n(x) = (-1)^n\,e^{x^2}\,\frac{d^n}{dx^n}\,e^{-x^2}\,, \qquad (n = 0, 1, 2, \ldots)\,, \tag{3.7.1}$$

from which we may deduce

$$H_{n+1}(x) = 2\,x\,H_n(x) - H_n'(x)\,. \tag{3.7.2}$$

In fact we have

$$H_{n+1}(x) = (-1)^{n+1}\,e^{x^2}\,\frac{d^{n+1}}{dx^{n+1}}\,e^{-x^2} = (-1)^{n+1}\,e^{x^2}\,\frac{d}{dx}\,[(-1)^n\,e^{-x^2}\,H_n(x)]$$

$$= -\,e^{x^2}\,[-\,2\,x\,e^{-x^2}\,H_n(x) + e^{-x^2}\,H_n'(x)]\,.$$

From the symmetry property and from (3.7.2) it is easy to deduce that if we put

$$H_n(x) = a_n\,x^n + b_n\,x^{n-1} + \cdots\,, \tag{3.7.3}$$

we have

$$a_n = 2^n\,, \qquad b_n = 0\,. \tag{3.7.4}$$

We prove now the orthogonality property of the polynomials determined by (3.7.1), that is that

$$\int_{-\infty}^{+\infty} e^{-x^2}\,H_m(x)\,H_n(x)\,dx = 0\,, \qquad (m < n)\,. \tag{3.7.5}$$

In fact, expressing $H_n(x)$ by means of (3.7.1) and operating m integrations by parts, the integral shown above becomes

$$(-1)^{n+m}\int_{-\infty}^{+\infty}\frac{d^m}{dx^m}H_m(x)\frac{d^{n-m}}{dx^{n-m}}e^{-x^2}\,dx=(-1)^{n+m}\,m!\,a_m\left[\frac{d^{n-m-1}}{dx^{n-m-1}}e^{-x^2}\right]_{-\infty}^{+\infty}=0.$$

If on the contrary $m = n$ the same procedure yields

$$\int_{-\infty}^{+\infty} e^{-x^2}\,H_n^2(x)\,dx = n!\,a_n\int_{-\infty}^{+\infty} e^{-x^2}\,dx = 2^n\,n!\,\sqrt{\pi}\,, \tag{3.7.6}$$

and therefore *the system*

$$H_n(x)\Big/\sqrt{2^n\,n!\,\sqrt{\pi}}\,,\qquad (n = 0, 1, 2, \ldots)\,, \tag{3.7.7}$$

is orthonormal in $(-\infty, +\infty)$ *with respect to the weight function* e^{-x^2}. Applying to this system the *Christoffel-Darboux formula* (*3.3.15*) we have

$$\sum_{k=0}^{n}\frac{H_k(x)\,H_k(y)}{2^k\,k!}=\frac{1}{2^{n+1}\,n!}\,\frac{H_{n+1}(x)\,H_n(y)-H_{n+1}(y)\,H_n(x)}{x-y}. \tag{3.7.8}$$

In order to obtain the *differential equation satisfied by the Hermite polynomials* it is sufficient to apply theor. 3.3.V remarking that (3.3.20) is now written $\dfrac{p'(x)}{p(x)} = -\,2\,x$ whence we have $A = 1$, $B = 0$, $C = 0$, $D = 0$, $E = -\,2$. Therefore (3.3.21) is written

$$y'' - 2\,x\,y' + 2\,n\,y = 0\,. \tag{3.7.9}$$

The usual *recurrence formula* of the type (3.3.10) can be soon obtained keeping into account that (3.3.11), (3.3.13), (3.3.14) together with (3.7.4) and (3.7.6) yield

$$A_n = \frac{1}{2}\,,\qquad B_n = 0\,,\qquad C_n = n\,,$$

whence (3.3.10) becomes

$$H_{n+1}(x) = 2\,x\,H_n(x) - 2\,n\,H_{n-1}(x) \tag{3.7.10}$$

and this formula, together with $H_0(x) = 1$, $H_1(x) = 2\,x$ allow the computation of all Hermite polynomials.

Finally note that from (3.7.2), (3.7.10) there immediately follows

$$H_n'(x) = 2\,n\,H_{n-1}(x)\,. \tag{3.7.11}$$

3.8. Some notions on divided differences. Let $f(x)$ be a function defined in an interval $[a, b]$. Having fixed in $[a, b]$ the points $x_1, x_2, \ldots, x_m$ $[x_i \neq x_j$ when $i \neq j]$, we call *divided difference* of $(m-1)$-th order of $f(x)$, relative to such points, the quantity defined by

$$f(x_1, x_2, \dots, x_m) = \frac{1}{V(x_1, x_2, \dots, x_m)} \begin{vmatrix} 1 & 1 & \dots & 1 \\ x_1 & x_2 & \dots & x_m \\ x_1^2 & x_2^2 & \dots & x_m^2 \\ \cdot & \cdot & \cdot & \cdot \\ x_1^{m-2} & x_2^{m-2} & \dots & x_m^{m-2} \\ f(x_1) & f(x_2) & \dots & f(x_m) \end{vmatrix}, \quad (3.8.1)$$

where $V(x_1, x_2, \dots, x_m)$ denotes the Vandermonde determinant of m numbers $x_1, x_2, \dots, x_m$. Expanding the determinant with respect to the elements of the last row, we immediately obtain

$$f(x_1, x_2, \dots, x_m) = \frac{f(x_1)}{(x_1 - x_2)(x_1 - x_3)\dots(x_1 - x_m)} + \frac{f(x_2)}{(x_2 - x_1)(x_2 - x_3)\dots(x_2 - x_m)} + \dots + \frac{f(x_m)}{(x_m - x_1)(x_m - x_2)\dots(x_m - x_{m-1})}. \quad (3.8.2)$$

Note also the following recurrence formula

$$f(x_1, x_2, \dots, x_m) = \frac{f(x_1, x_2, \dots, x_{m-2}, x_m) - f(x_1, x_2, \dots, x_{m-2}, x_{m-1})}{x_m - x_{m-1}} \quad (3.8.3)$$

which can be easily proved expressing the first and the second member by means of (3.8.2).

Having fixed the points $x_1, x_2, \dots, x_m$ let us consider the following divided difference of m-th order

$$f(x_1, x_2, \dots, x_m, x)$$

which comes out tó be a function of x defined on the set obtained from the interval $[a, b]$ taking off the points $x_1, x_2, \dots, x_m$. Using (3.8.1) and De l'Hospital theorem it is soon that, if $f(x) \in C^1[a, b]$, the limits of $f(x_1, x_2, \dots, \dots, x_m, x)$ for $x \to x_i$, $(i = 1, 2, \dots, m)$, are determined and finite, so that $f(x_1, x_2, \dots, x_m, x)$ appears continuous in all $[a, b]$. Analogously we see that, if $f(x) \in C^k[a, b]$, $(k \geqq 1)$, we have then $f(x_1, x_2, \dots, x_m, x) \in C^{k-1}[a, b]$.

We now prove the following theorems:

Theorem 3.8.I. *Under the hypothesis that $f(x) \in C^m[a, b]$ we may associate to every point $x \in [a, b]$ at least one point $\xi \in [a, b]$ in such a way as to have*

$$f(x_1, x_2, \dots, x_m, x) = \frac{f^{(m)}(\xi)}{m!}. \quad (3.8.4)$$

Proof. Having considered the function

$$F(t) = \begin{vmatrix} 1 & 1 & \dots & 1 & 1 & 1 \\ x_1 & x_2 & \dots & x_m & x & t \\ x_1^2 & x_2^2 & \dots & x_m^2 & x^2 & t^2 \\ \cdot & \cdot & \cdot & \cdot & \cdot & \cdot \\ x_1^m & x_2^m & \dots & x_m^m & x^m & t^m \\ f(x_1) & f(x_2) & \dots & f(x_m) & f(x) & f(t) \end{vmatrix},$$

we evidently have $F(x_1) = F(x_2) = \cdots = F(x_m) = F(x) = 0$ and therefore applying m times the Rolle theorem we may conclude that there exists at least one point $\xi \in [a, b]$ such that we have

$$F^{(m)}(\xi) = \begin{vmatrix} 1 & 1 & \dots 1 & 1 & 0 \\ x_1 & x_2 & \dots x_m & x & 0 \\ x_1^2 & x_2^2 & \dots x_m^2 & x^2 & 0 \\ \cdot & \cdot & \cdot & \cdot & \cdot \\ x_1^m & x_2^m & \dots x_m^m & x^m & m! \\ f(x_1) & f(x_2) & \dots f(x_m) & f(x) & f^{(m)}(\xi) \end{vmatrix} = 0 .$$

Expanding this determinant with respect to the elements of the last column, we obtain

$$-m! \begin{vmatrix} 1 & 1 & \dots 1 & 1 \\ x_1 & x_2 & \dots x_m & x \\ \cdot & \cdot & \cdot & \cdot \\ x_1^{m-1} & x_2^{m-1} & \dots x_m^{m-1} & x^{m-1} \\ f(x_1) & f(x_2) & \dots f(x_m) & f(x) \end{vmatrix} + f^{(m)}(\xi)\, V(x_1, x_2, \dots, x_m, x) = 0$$

and this, for (3.8.1), is equivalent to (3.8.4), q.e.d.

Theorem 3.8.II. *Under the hypothesis that $f(x) \in C^{m+1}[a, b]$ we may associate to every point $x \in [a, b]$ at least one point $\xi \in [a, b]$ in such a way that we have*

$$\frac{d}{dx} f(x_1, x_2, \dots, x_m, x) = \frac{f^{(m+1)}(\xi)}{(m+1)!} . \tag{3.8.5}$$

Proof. In virtue of the preceding theorem, we may write

$$f(x_1, x_2, \dots, x_m, x, x+h) = \frac{f^{(m+1)}(\xi_h)}{(m+1)!} ,$$

where ξ_h denotes a certain point of the interval $[a, b]$ and then, expressing the first member by means of (3.8.3):

$$\frac{f(x_1, x_2, \dots, x_m, x+h) - f(x_1, x_2, \dots, x_m, x)}{h} = \frac{f^{(m+1)}(\xi_h)}{(m+1)!} .$$

For $h \to 0$ the first member tends to $\frac{d}{dx} f(x_1, x_2, \dots, x_m, x)$; the second member tends to the same limit, which must evidently coincide with one of the value taken by the continuous function $\frac{f^{(m+1)}(x)}{(m+1)!}$ in $[a, b]$. Therefore there exists at least one point $\xi \in [a, b]$ in which this function is equal to the mentioned limit; from it there follows (3.8.5) q.e.d.

Theorem 3.8.III. *Under the hypothesis $f(x) \in C^{m+2}[a, b]$ we may associate to every point $x \in [a, b]$ at least one point $\xi \in [a, b]$ in such a way as to have*

$$\frac{1}{2} \frac{d^2}{dx^2} f(x_1, x_2, \dots, x_m, x) = \frac{f^{(m+2)}(\xi)}{(m+2)!} \tag{3.8.6}$$

Proof. The proof is analogous to the preceding one, starting from

$$f(x_1, x_2, \ldots, x_m, x, x + h, x + 2h) = \frac{f^{(m+2)}(\xi_h)}{(m+2)!}$$

which [applying twice (3.8.3)] is equivalent to

$$\frac{f(x_1, \ldots, x_m, x + 2h) - 2f(x_1, \ldots, x_m, x + h) + f(x_1, \ldots, x_m, x)}{2h^2} = \frac{f^{(m+2)}(\xi_h)}{(m+2)!}.$$

3.9. The s-orthogonal polynomials connected to a given weight function. In a finite or infinite interval $[a, b]$ let be defined a weight function $p(x)$, measurable and non negative, being $p(x) > 0$ in a set of positive measure and $p(x)\, x^n \in L[a, b]$, $(n = 0, 1, 2, \ldots)$.[1]) Let then a sequence $\{a_n\}_{n=0,1,2,\ldots}$ of numbers *all different from zero* be assigned.

Having fixed an integer $s \geqq 0$, we want to construct a sequence $\{P_{s,n}(x)\}_{n=0,1,2,\ldots}$, *where $P_{s,n}(x)$ is a polynomial of n degree having a_n as coefficient of x^n, in such a way that*

$$\int_a^b p(x)\, P_{s,m}(x)\, [P_{s,n}(x)]^{2s+1}\, dx = 0 \quad \text{if } m < n, \tag{3.9.1}$$

or, under an equivalent form

$$\int_a^b p(x)\, \Pi_{n-1}(x)\, [P_{s,n}(x)]^{2s+1}\, dx = 0, \quad (n = 1, 2, \ldots), \tag{3.9.2}$$

where $\Pi_{n-1}(x)$ denotes an arbitrary polynomial of degree $\leqq n - 1$. It is evident that if $s = 0$ the problem coincides with that of the orthogonal polynomials considered in § 3.3; on the contrary for $s > 0$ we have a new problem and the polynomials $P_{s,n}(x)$ (if they exist) will be called *s-orthogonal* in *the interval* $[a, b]$ *with respect to the weight function* $p(x)$.

Let us begin by proving the following theorems:

Theorem 3.9.I. *If there exists a sequence of polynomials $\{P_{s,n}(x)\}_{n=0,1,2,\ldots}$ satisfying the above mentioned conditions, this is necessarily unique.*

Proof. In fact, should exist another sequence $\{Q_{s,n}(x)\}_{n=0,1,2,\ldots}$ with $Q_{s,n}(x) = a_n x^n + \cdots$, the polynomial $P_{s,n}(x) - Q_{s,n}(x)$ would be of degree $\leqq n - 1$ and therefore, using (3.9.2), there would hold

$$\int_a^b p(x)\, [P_{s,n}(x) - Q_{s,n}(x)]\, [P_{s,n}(x)]^{2s+1}\, dx = 0,$$

$$\int_a^b p(x)\, [P_{s,n}(x) - Q_{s,n}(x)]\, [Q_{s,n}(x)]^{2s+1}\, dx = 0,$$

which, with a subtraction member by member, give

$$\int_a^b p(x)\, [P_{s,n}(x) - Q_{s,n}(x)]\, \{[P_{s,n}(x)]^{2s+1} - [Q_{s,n}(x)]^{2s+1}\}\, dx = 0,$$

[1]) These conditions of summability reduce to $p(x) \in L[a, b]$ if $[a, b]$ is finite.

that is

$$\int_a^b p(x)\,[P_{s,n}(x) - Q_{s,n}(x)]^2 \sum_{\nu=0}^{2s} [P_{s,n}(x)]^{2s-\nu}\,[Q_{s,n}(x)]^\nu\,dx = 0\,,$$
$$(n = 1, 2, \ldots)\,. \tag{3.9.3}$$

In the integrand, the first factor is >0 in a set of positive measure, the second factor is $\geqq 0$, the third factor is >0 except at most in a finite number of points and therefore (3.9.3) may hold only if the second factor is identically zero, that is if $Q_{s,n}(x) \equiv P_{s,n}(x)$, q.e.d.

From this theorem we can obviously deduce that if we do not preassign the values a_n of the coefficients of x^n, namely, *if we give only the condition of s-orthogonality (3.9.1), then every polynomial $P_{s,n}(x)$ (if it exists) appears determined apart from an arbitrary constant factor.*

Theorem 3.9.II. *If there exists a sequence of polynomials $\{P_{s,n}(x)\}_{n=0,1,2,\ldots}$ satisfying the above mentioned conditions, then the zeros of every polynomial $P_{s,n}(x)$, $(n = 1, 2, \ldots)$, are real, distinct and internal in the interval $[a, b]$.*

Proof. It is quite analogous to that of theorem 3.3.III.

Theorem 3.9.III. *Let the interval $[a, b]$ be of the type $[-c, c]$ and the weight $p(x)$ be an even function. Then the polynomials (if they exist) are even or odd functions, according to whether n is even or odd.*

Proof. It is quite analogous to that of theorem 3.3.IV.

We shall now prove the existence of the polynomials $P_{s,n}(x)$ (see I. Verna [1]) basing ourselves on the following classical theorem of functional analysis: *if S is a linear*[1]) *and normed space, fixing arbitrarily n of its points $u_0, u_1, \ldots, u_{n-1}$ linearly independent, to every point $u \in S$ we make correspond at least a linear combination $\sum_{k=0}^{n-1} c_k u_k$ of the points $u_0, u_1, \ldots, u_{n-1}$ which be of best approximation for the point u, that is such as to give*

$$\left\|u - \sum_{k=0}^{n-1} c_k u_k\right\| = \text{minimum.}^{2)} \tag{3.9.4}$$

The space S that we want to consider here is the space $L^{2s+2}[a, b]$ of the functions $u = u(x)$ measurable in $[a, b]$ for which we have $\int_a^b [u(x)]^{2s+2}\,dx < +\infty$; it is well known that such a space is linear and normed when we put

$$\|u\| = \left(\int_a^b [u(x)]^{2s+2}\,dx\right)^{\frac{1}{2s+2}}. \tag{3.9.5}$$

[1]) Here it suffices to assume that S is linear in the real domain.
[2]) See for instance N. I. Achieser [1], § 8. The symbol $\|u\|$ denotes the norm of u in the space S.

Having fixed in $L^{2s+2}[a, b]$ the n points $u_k = x^k[p(x)]^{\frac{1}{2s+2}}$, $(k = 0, 1, \ldots, n-1)$, and having put $u = a_n x^n[p(x)]^{\frac{1}{2s+2}}$, the above mentioned theorem gives us the certainty that there exist n constants $c_0, c_1, \ldots, c_{n-1}$ such as to give

$$\left\| u - \sum_{k=0}^{n-1} c_k u_k \right\| = \left(\int_a^b p(x) \left[a_n x^n - \sum_{k=0}^{n-1} c_k x^k \right]^{2s+2} dx \right)^{\frac{1}{2s+2}} = \text{minimum} .$$

In other words, for every n and for every choice of $a_n \neq 0$, there exists a polynomial of degree n:

$$P_{s,n}(x) = a_n x^n - \sum_{k=0}^{n-1} c_k x^k$$

such to yield

$$\int_a^b p(x) \, [P_{s,n}(x)]^{2s+2} \, dx = \text{minimum} . \tag{3.9.6}$$

There follows that each of the n functions

$$F_k(\lambda) = \int_a^b p(x) \, [P_{s,n}(x) + \lambda x^k]^{2s+2} \, dx , \qquad (k = 0, 1, \ldots, n-1) ,$$

must have its derivative null for $\lambda = 0$. But we have

$$F_k'(\lambda) = (2s+2) \int_a^b p(x) \, x^k \, [P_{s,n}(x) + \lambda x^k]^{2s+1} \, dx$$

and therefore $F_k'(0) = 0$ give

$$\int_a^b p(x) \, x^k \, [P_{s,n}(x)]^{2s+1} \, dx = 0 , \qquad (k = 0, 1, \ldots, n-1) ,$$

which express that the polynomial $P_{s,n}(x)$ satisfies (3.9.2). We have thus proved, for every n, the existence of $P_{s,n}(x) = a_n x^n + \cdots$ satisfying the wanted conditions.[1])

Note that, *in the case that* $[a, b]$ *be infinite*, the formula (3.9.6) shows that to prove only the existence of the polynomials $P_{s,n}(x)$ with n preassigned, it suffices to make on the weight $p(x)$ the only hypothesis

$$p(x) \, x^{n(2s+2)} \in L[a, b] . \tag{3.9.7}$$

The theory of s-orthogonal polynomials (with $s > 0$) has not yet been deeply studied.[2]) We shall therefore confine ourselves to giving only some indications in the following particular cases

$$a = -1 , \quad b = 1 , \quad p(x) = 1 ; \tag{3.9.8}$$

$$a = 0 , \quad b = +\infty , \quad p(x) = e^{-x} ; \tag{3.9.9}$$

$$a = -\infty , \quad b = +\infty , \quad p(x) = e^{-x^2} . \tag{3.9.10}$$

[1]) For another more elementary proof, which is valid only if the interval $[a, b]$ is finite, see § 3.10, Problem 11.

[2]) See for instance P. TURAN [1], T. POPOVICIU [1], A. GHIZZETTI-A. OSSICINI [1].

In the case (3.9.8) we may apply theorem 3.9.III and deduce that the zeros of $P_{s,n}(x)$ are set symmetrically with respect to the origin. We give here a table where are shown the *positive* zeros of $P_{s,n}(x)$ for $s = 1, 2$ and $n = 2, 3, 4, 5$. [1])

	$n = 2$	$n = 3$	$n = 4$	$n = 5$
$s = 1$	0,6292111	0,8144392	0,3588585 0,8896768	0,5608674 0,9271179
$s = 2$	0,6500278	0,7960777	0,3659244 0,8998292	0,5690324 0,9343690

In this case it is convenient to choose the factor or proportionality a_n in such a way as to have $P_{s,n}(1) = 1$; with that for $s = 0$ we find the Legendre polynomials $P_n(x)$. [2]) We prove that *if there holds*

$$\int_{-1}^{1} [P_{s,n}(x)]^{2s+2}\, dx = \frac{2}{1 + (2s + 2)\, n}, \tag{3.9.11}$$

which generalizes (3.5.6) valid for the Legendre polynomials. In fact integrating by parts we get

$$\int_{-1}^{1} [P_{s,n}(x)]^{2s+2}\, dx = \{x[P_{s,n}(x)]^{2s+2}\}_{-1}^{1} - (2s + 2) \int_{-1}^{1} x\, [P_{s,n}(x)]^{2s+1}\, P'_{s,n}(x)\, dx\,. \tag{3.9.12}$$

Keeping into account that $x\, [P_{s,n}(x)]^{2s+2}$ is an odd function with value 1 for $x = 1$ and that we have $x\, P'_{s,n}(x) = n\, P_{s,n}(x) + \Pi_{n-2}(x)$ with $\Pi_{n-2}(x)$ polynomial of degree $n - 2$, (3.9.12) becomes

$$\int_{-1}^{1} [P_{s,n}(x)]^{2s+2}\, dx = 2 - (2s + 2) \int_{-1}^{1} [P_{s,n}(x)]^{2s+1}\, [n\, P_{s,n}(x) + \Pi_{n-2}(x)]\, dx$$

and therefore for (3.9.2)

$$\int_{-1}^{1} [P_{s,n}(x)]^{2s+2}\, dx = 2 - (2s + 2)\, n \int_{-1}^{1} [P_{s,n}(x)]^{2s+2}\, dx;$$

this formula is equivalent to (3.9.11).

In the case (3.9.9) we give only a table with the zeros of $P_{s,n}(x)$ for $s = 1$ and $n = 1, 2, 3$. [3])

[1]) See L. Rebolia [1].

[2]) That it is $P_n(1) = 1$, $P_n(-1) = (-1)^n$ follows from (3.4.4) and (3.4.5) putting $\alpha = \beta = 0$.

[3]) See I. Verna [1].

	$n = 1$	$n = 2$	$n = 3$
$s = 1$	1,5960716	0,8934994 6,3877983	0,6212774 4,1936031 12,1938097

In the case (3.9.10) we have still the symmetry of the zeros of $P_{s,n}(x)$ with respect to the origin. Here it is a table of the *positive* zeros of $P_{s,n}(x)$ for $s = 1, 2$ and $n = 2, 3$. [1])

	$n = 2$	$n = 3$
$s = 1$	1,0264374	1,7699914
$s = 2$	1,2686013	2,1841838

3.10. Problems.

1. Prove that for $|z| < \pi$ we have

$$\tanh\frac{z}{2} = \sum_{s=0}^{\infty} c_s \frac{z^{2s+1}}{(2s+1)!} \quad \text{with} \quad c_s = \frac{2^{2s+2}-1}{s+1} B_{2s+2}\,. \tag{1}$$

Deduce that the coefficients c_s have the properties expressed by

$$\sum_{s=0}^{p-1} c_s \binom{2p}{2s+1} = 1\,, \quad (p = 1, 2, \ldots)\,, \tag{2}$$

$$\sum_{s=0}^{p-1} c_s \binom{2p+1}{2s+1} + 2c_p = 1\,, \quad (p = 0, 1, 2, \ldots)\,. \tag{3}$$

2. Prove that we have

$$z \operatorname{cotan} z = \sum_{n=0}^{\infty} (-1)^n 2^{2n} B_{2n} \frac{z^{2n}}{(2n)!}\,, \quad (|z| < \pi)\,, \tag{1}$$

$$\tan z = \sum_{n=1}^{\infty} (-1)^{n-1} 2^{2n} (2^{2n}-1) B_{2n} \frac{z^{2n-1}}{(2n)!}\,, \quad \left(|z| < \frac{\pi}{2}\right). \tag{2}$$

3. Prove that we have

$$B_n(x) = \sum_{k=0}^{n} \binom{n}{k} B_{n-k}\, x^k\,. \tag{1}$$

4. Prove that, if m is a positive integer, we have

$$B_n(mx) = m^{n-1} \sum_{k=0}^{m-1} B_n\left(x + \frac{k}{m}\right). \tag{1}$$

5. Prove that $\lim_{\alpha \to 0} \Gamma(\alpha) = +\infty$.

[1]) See I. VERNA [1].

6. Prove that the Gauss formula holds

$$\Gamma(\alpha) = \lim_{n\to\infty} \frac{n^\alpha\, n!}{\alpha(\alpha+1)(\alpha+2)\ldots(\alpha+n)}, \quad (\alpha > 0). \tag{1}$$

7. We can obtain the extension of the function $\Gamma(\alpha)$ to the *non integer* negative values of α by means of the following formula [suggested by (3.2.5)]:

$$\Gamma(\alpha) = \frac{\Gamma(\alpha+n)}{\alpha(\alpha+1)\ldots(\alpha+n-1)}, \tag{1}$$

where n denotes an integer for which we get $\alpha + n > 0$. Prove that for k integer non negative we have

$$\lim_{\alpha\to -k} (\alpha + k)\,\Gamma(\alpha) = \frac{(-1)^k}{k}. \tag{2}$$

8. Prove that

$$\Gamma(x+\alpha) \leqq x^\alpha\, \Gamma(x), \quad (x > 0,\ 0 \leq \alpha \leq 1). \tag{1}$$

9. Prove that for the zeros $x_{ni}^{(\alpha,\beta)}$, $(i = 1, 2, \ldots, n)$, of the Jacobi polynomial $P_n^{(\alpha,\beta)}(x)$ there holds

$$\sum_{i=1}^{n} x_{ni}^{(\alpha,\beta)} = \frac{n(\beta-\alpha)}{\alpha+\beta+2n}. \tag{1}$$

10. Prove that for the zeros $x_{ni}^{(\alpha,\beta)}$, $(i = 1, 2, \ldots, n)$, of the Jacobi polynomial $P_n^{(\alpha,\beta)}(x)$ there holds

$$\lim_{\alpha\to+\infty} x_{ni}^{(\alpha,\beta)} = -1, \quad (n, \beta \text{ fixed}); \quad \lim_{\beta\to+\infty} x_{ni}^{(\alpha,\beta)} = 1, \quad (\alpha, n \text{ fixed}). \tag{1}$$

11. To prove the existence and the uniqueness (apart from the constant factor a_n) of the polynomials $P_{s,n}(x)$ which are s-orthogonal in $[a, b]$ with respect to the weight $p(x) \geqq 0$, we may take as unknowns the zeros $x_{n1}, x_{n2}, \ldots, x_{nn}$ of $P_{s,n}(x)$ $(n \geqq 1)$ and write that they must satisfy the following n equations [see (3.9.2)]:

$$\varphi_k(x_{n1}, x_{n2}, \ldots, x_{nn}) = \int_a^b p(x) \cdot \prod_{i=1}^{k} (x - x_{ni}) \cdot \prod_{j=1}^{n} (x - x_{nj})^{2s+1}\, dx = 0,$$
$$(k = 0, 1, \ldots, n-1).\ ^{1)} \tag{1}$$

Assuming that $[a, b]$ be *finite*, prove, by induction, that the system (1) has one and only one solution $x_{n1}, x_{n2}, \ldots, x_{nn}$ satisfying the condition

$$a < x_{n1} < x_{n2} < \cdots < x_{nn} < b. \tag{2}$$

[1]) It is clear that $\prod_{i=1}^{k} (x - x_{ni})$ must be substituted by 1 when $k = 0$.

CHAPTER 4

Various examples of elementary quadrature formulae

4.1. General remarks on the contents of this chapter. In this chapter we give numerous examples of elementary quadrature formulae obtained by applying the general rules given in chapter 2. We state again the three fundamental formulae:[1])

$$\int_a^b g(x)\, u(x)\, dx = \sum_{h=0}^{n-1} \sum_{i=1}^{m} A_{hi}\, u^{(h)}(x_i) + R(u)\,, \tag{4.1.1}$$

$$A_{hi} = [E^*_{n-h-1}(\varphi_i - \varphi_{i-1})]_{x=x_i}\,, \tag{4.1.2}$$

$$R(u) = \sum_{i=0}^{m} \int_{x_i}^{x_{i+1}} \varphi_i(x)\, E[u(x)]\, dx = \int_a^b \Phi(x)\, E[u(x)]\, dx\,, \tag{4.1.3}$$

where the functions $\varphi_i(x)$, $(i = 0, 1, \ldots, m-1, m)$ are *arbitrary* solutions of the differential equation $E^*(\varphi) = g$, with the exception of $\varphi_0(x)$ and $\varphi_m(x)$ which must satisfy respectively the initial conditions null at the point a and at the point b. Remember that $\varphi_0(x)$ need not be considered if $x_1 = a$, nor $\varphi_m(x)$ if $x_m = b$.

In certain examples we preassign the weight $g(x)$, the nodes $x_1, \ldots, x_m$, the operator E of order n *and also the solutions* $\varphi_1(x), \ldots, \varphi_{m-1}(x)$; we have only to apply the preceding formulae [see § 4.2 and 4.3].

In other examples, having fixed the weight, the nodes and the operator E, we leave indetermined the choice of the solutions $\varphi_1(x), \ldots, \varphi_{m-1}(x)$ and we use the $(m-1)\,n$ arbitrary parameters which thus occur in the quadrature formula to apply some other condition [see § 4.4].

Other examples are presented as *Gauss problems* [see § 4,5, . . ., 4.11 and 4.13, 4.14] or as *Tchebychef problems* [see § 4.12, 4.14.].

In all examples the hypotheses on the argument function $u(x)$ are specified and from the expression (4.1.3) of the remainder we deduce some formulae of bounding or evaluation of the remainder itself.

4.2. Deduction of three classical formulae. In this § we assume a, b *finite*. Let us begin by deducing the so called *formula of the circumscribed trapezium*, assuming

$$g(x) = 1; \quad m = 1\,, \quad x_1 = \frac{a+b}{2}; \quad n = 2\,, \quad E = \frac{d^2}{dx^2}.$$

[1]) See (2.1.4), (2.1.12), (2.1.13) (for a, b finite); (2.2.6), (2.2.14), (2.2.15) (for a finite, $b = +\infty$); (2.3.6), (2.3.12), (2.3.13) (for $a = -\infty$, $b = +\infty$). For the influence-function $\Phi(x)$ see (2.1.18).

Since $m = 1$ there is no freedom of choice of the solutions $\varphi_i(x)$ of the differential equation $E^*(\varphi) = g$ which now must be written $\varphi'' = 1$. We have only to consider the fully determined solutions $\varphi_0(x)$, $\varphi_1(x)$ for which we evidently have

$$\varphi_0(x) = \frac{1}{2}(x-a)^2, \qquad \varphi_1(x) = \frac{1}{2}(x-b)^2.$$

Then, considering that $E_0^* = 1$, $E_1^* = -\frac{d}{dx}$, equation (4.1.2) immediately yields

$$A_{01} = b - a, \qquad A_{11} = 0,$$

so that, under the hypothesis $u(x) \in A\,C^1[a, b]$, (4.1.1) becomes

$$\boxed{\int_a^b u(x)\,dx = (b-a)\,u\left(\frac{a+b}{2}\right) + R(u)}, \tag{4.2.1}$$

with

$$R(u) = \int_a^b \Phi(x)\,u''(x)\,dx; \qquad \Phi(x) = \begin{cases} \frac{1}{2}(x-a)^2, & \left(a \leqq x \leqq \frac{a+b}{2}\right), \\ \frac{1}{2}(x-b)^2, & \left(\frac{a+b}{2} \leqq x \leqq b\right). \end{cases} \tag{4.2.2}$$

The influence function $\Phi(x)$ is positive for $a < x < b$ and has the graph of Fig. 1; therefore from (4.2.2) we get

$$|R(u)| \leqq \frac{(b-a)^2}{8} \int_a^b |u''(x)|\,dx = \frac{(b-a)^2}{8} V_1 \tag{4.2.3}$$

where V_1 denotes the *total variation* of the function $u'(x)$ absolutely continuous on the interval $[a, b]$.

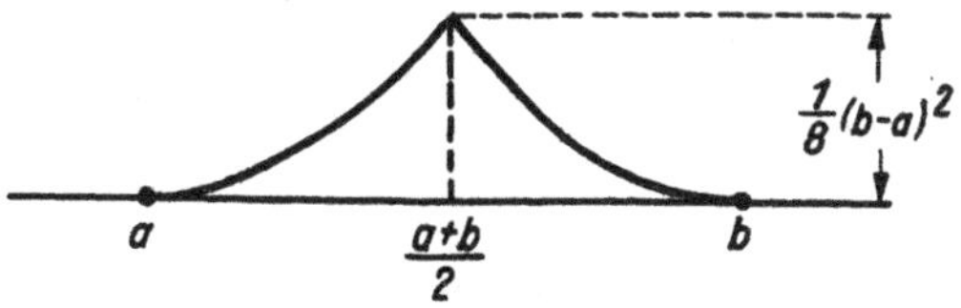

Fig. 1

Under a more restrictive hypothesis that $u''(x)$ *be bounded in* $[a, b]$ with $M_2 = \sup_{a \leqq x \leqq b} |u''(x)|$, we may deduce from (4.2.2)

$$|R(u)| \leqq M_2 \int_a^b \Phi(x)\,dx = \frac{(b-a)^3}{24} M_2. \tag{4.2.4}$$

Moreover if we assume that $u''(x)$ *be continuous in* $[a, b]$ and use the fact that $\Phi(x)$ is continuous and does not change sign, we may apply the mean value theorem and write

$$R(u) = u''(\xi) \int_a^b \Phi(x)\, dx = \frac{(b-a)^3}{24} u''(\xi)\,, \quad (a < \xi < b)\,. \tag{4.2.5}$$

Now we obtain the *formula of the inscribed trapezium* assuming

$$g(x) = 1; \quad m = 2\,, \quad x_1 = a\,, \quad x_2 = b; \quad n = 2\,, \quad E = \frac{d^2}{dx^2}\,.$$

In this case we need not consider the solutions $\varphi_0(x), \varphi_2(x)$ of the differential equation $\varphi'' = 1$ and we may choose arbitrarily the solution $\varphi_1(x)$; let us assume

$$\varphi_1(x) = \frac{1}{2}(x-a)(x-b)\,.$$

Considering that $E_0^* = 1$, $E_1^* = -\frac{d}{dx}$, (4.1.2) immediately yields

$$A_{01} = A_{02} = \frac{b-a}{2}\,, \qquad A_{11} = A_{12} = 0\,,$$

whence, *under the hypothesis* $u(x) \in A\ C^1[a, b]$, (4.1.1) gives

$$\boxed{\int_a^b u(x)\, dx = \frac{b-a}{2}[u(a) + u(b)] + R(u)}\,, \tag{4.2.6}$$

with

$$R(u) = \int_a^b \Phi(x)\, u''(x)\, dx; \qquad \Phi(x) = \frac{1}{2}(x-a)(x-b)\,. \tag{4.2.7}$$

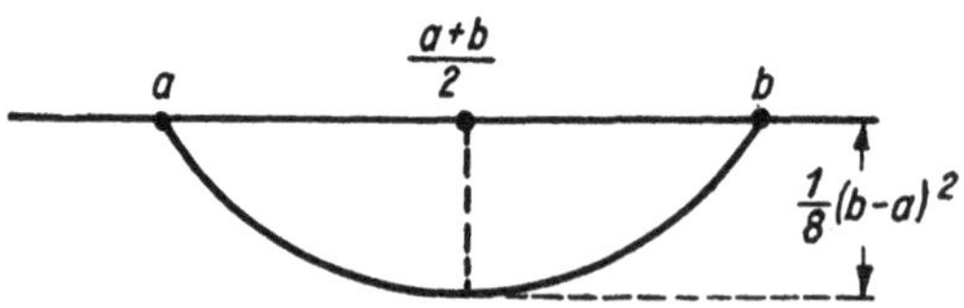

Fig. 2

The influence function $\Phi(x)$ is negative for $a < x < b$ and has the graph of Fig. 2; therefore, with the notations already introduced in the preceding case, we obtain

$$|R(u)| \leqq \frac{(b-a)^2}{8} V_1\,, \tag{4.2.8}$$

or, assuming $u''(x)$ *bounded in* $[a, b]$:

$$|R(u)| \leqq M_2 \int_a^b |\Phi(x)|\, dx = \frac{(b-a)^3}{12} M_2\,. \tag{4.2.9}$$

If moreover $u''(u)$ *is continuous in* $[a, b]$ *we have*

$$R(u) = u''(\xi) \int_a^b \Phi(x)\, dx = -\frac{(b-a)^3}{12} u''(\xi) . \tag{4.2.10}$$

Finally we obtain the *Cavalieri-Simpson formula*, assuming

$$g(x) = 1; \quad m = 3, \quad x_1 = a, \quad x_2 = \frac{a+b}{2}, \quad x_3 = b; \quad n = 4, \quad E = \frac{d^4}{dx^4} .$$

The differential equation $E^*(\varphi) = g$ is to be written $\varphi'''' = 1$. We are not interested in its solution $\varphi_0(x)$, $\varphi_3(x)$; we may fix arbitrarily the two integrals $\varphi_1(x)$, $\varphi_2(x)$, so we shall assume

$$\varphi_1(x) = \frac{1}{24}(x-a)^3\left(x - \frac{a+2b}{3}\right), \quad \varphi_2(x) = \frac{1}{24}(x-b)^3\left(x - \frac{2a+b}{3}\right).$$

Since $E_0^* = 1$, $E_1^* = -\frac{d}{dx}$, $E_2^* = \frac{d^2}{dx^2}$, $E_3^* = -\frac{d^3}{dx^3}$, (4.1.2) gives, with an easy calculation

$$A_{01} = \frac{b-a}{6}, \quad A_{02} = 2\frac{b-a}{3}, \quad A_{03} = \frac{b-a}{6},$$

$$A_{11} = A_{12} = A_{13} = 0, \quad A_{21} = A_{22} = A_{23} = 0, \quad A_{31} = A_{32} = A_{33} = 0$$

whence we have the formula, under the hypothesis $u(x) \in A\, C^3[a, b]$:

$$\boxed{\int_a^b u(x)\, dx = \frac{b-a}{6}\left[u(a) + 4\, u\left(\frac{a+b}{2}\right) + u(b)\right] + R(u)}, \tag{4.2.11}$$

with

$$\left.\begin{aligned} R(u) &= \int_a^b \Phi(x)\, u''''(x)\, dx; \\ \Phi(x) &= \begin{cases} \frac{1}{24}(x-a)^3\left(x - \frac{a+2b}{3}\right), & \left(a \leqq x \leqq \frac{a+b}{2}\right), \\ \frac{1}{24}(x-b)^3\left(x - \frac{2a+b}{3}\right), & \left(\frac{a+b}{2} \leqq x \leqq b\right). \end{cases} \end{aligned}\right\} \tag{4.2.12}$$

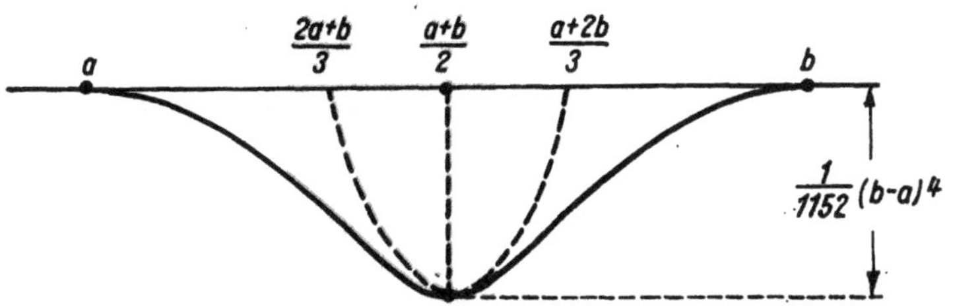

Fig. 3

The influence-function $\Phi(x)$ is negative for $a < x < b$ and has the graph of Fig. 3; therefore from (4.2.12) we get

$$|R(u)| \leq \frac{(b-a)^4}{1152} \int_a^b |u''''(x)|\, dx = \frac{(b-a)^4}{1152} V_3 , \tag{4.2.13}$$

where V_3 denotes the *total variation* of the function $u'''(x)$ which is absolutely continuous in the interval $[a, b]$. If we assume $u''''(x)$ *to be bounded in* $[a, b]$, namely $M_4 = \sup_{a \leq x \leq b} |u''''(x)|$, from (4.2.12) we may deduce:

$$|R(u)| \leq M_4 \int_a^b |\Phi(x)|\, dx = \frac{(b-a)^5}{2880} M_4 , \tag{4.2.14}$$

while under the more restrictive hypothesis that $u''''(x)$ *be continuous in* $[a, b]$, the application of the mean value theorem gives

$$R(u) = u''''(\xi) \int_a^b \Phi(x)\, dx = -\frac{(b-a)^5}{2880} u''''(\xi) , \quad (a < \xi < b) . \tag{4.2.15}$$

4.3. The Euler-MacLaurin formula. Let us apply the usual general rule assuming

$$a = 0 , \quad b = 1 ; \quad g(x) = 1 ; \quad m = 2 , \quad x_1 = 0 , \quad x_2 = 1 ;$$

$$n = 2\nu + 2 , \quad E = \frac{d^{2\nu+2}}{dx^{2\nu+2}} , \qquad (\nu = 0, 1, 2, \ldots) .$$

The differential equation $E^*(\varphi) = g$ is to be written $\varphi^{(2\nu+2)} = 1$. We need not consider the solutions $\varphi_0(x)$, $\varphi_2(x)$ and it sufficies to fix arbitrarily the solution $\varphi_1(x)$; for instance we may assume

$$\varphi_1(x) = \frac{B_{2\nu+2}(x) - B_{2\nu+2}}{(2\nu+2)!} ,$$

where $B_{2\nu+2}(x)$, $B_{2\nu+2}$ denote respectively the *Bernoulli polynomial and number* of index $2\nu + 2$ [see § 3.1; particularly (3.1.9) with $k = n$].

Equations (4.1.1), (4.1.3) then become, *under the hypothesis* $u(x) \in A\, C^{2\nu+1}[0, 1]$:

$$\int_0^1 u(x)\, dx = \sum_{h=0}^{2\nu+1} [A_{h1} u^{(h)}(0) + A_{h2} u^{(h)}(1)] + R(u) , \tag{4.3.1}$$

$$R(u) = \int_0^1 \frac{B_{2\nu+2}(x) - B_{2\nu+2}}{(2\nu+2)!} u^{(2\nu+2)}(x)\, dx , \tag{4.3.2}$$

and we have only to compute the coefficients A_{h1}, A_{h2} using (4.1.2). We find

$$A_{h1} = [E^*_{2\nu+2-h-1}(\varphi_1)]_{x=0} = (-1)^{2\nu+1-h}\,\varphi_1^{(2\nu+1-h)}(0)$$

$$= \begin{cases} (-1)^{h+1}\dfrac{B^{(2\nu+1-h)}_{2\nu+2}(0)}{(2\nu+2)!}, & (h = 0, 1, \ldots, 2\nu), \\ \dfrac{B_{2\nu+2}(0) - B_{2\nu+2}}{(2\nu+2)!}, & (h = 2\nu+1), \end{cases}$$

$$A_{h2} = [E^*_{2\nu+2-h-1}(-\varphi_1)]_{x=1} = -(-1)^{2\nu+1-h}\,\varphi_1^{(2\nu+1-h)}(1)$$

$$= \begin{cases} (-1)^{h}\dfrac{B^{(2\nu+1-h)}_{2\nu+2}(1)}{(2\nu+2)!}, & (h = 0, 1, \ldots, 2\nu), \\ -\dfrac{B_{2\nu+2}(1) - B_{2\nu+2}}{(2\nu+2)!}, & (h = 2\nu+1), \end{cases}$$

and therefore, recalling (3.1.2), (3.1.6), (3.1.10), (3.1.11):

$$A_{01} = \frac{1}{2}, \qquad A_{02} = \frac{1}{2},$$

$$A_{h1} = \begin{cases} 0, & (h \text{ even}) \\ \dfrac{B_{h+1}}{(h+1)!}, & (h \text{ odd}) \end{cases}, \qquad A_{h2} = \begin{cases} 0, & (h \text{ even}) \\ -\dfrac{B_{h+1}}{(h+1)!}, & (h \text{ odd}) \end{cases},$$

$$(h = 1, 2, \ldots, 2\nu),$$

$$A_{2\nu+1,1} = 0, \qquad A_{2\nu+1,2} = 0.$$

Thus (4.3.1) reduces to the classical *Euler-MacLaurin formula*:

$$\boxed{\int_0^1 u(x)\,dx = \frac{1}{2}[u(0) + u(1)] + \sum_{k=1}^{\nu}\frac{B_{2k}}{(2k)!}[u^{(2k-1)}(0) - u^{(2k-1)}(1)] + R(u), \quad (\nu = 0, 1, 2, \ldots),} \tag{4.3.3}$$

where for the remainder $R(u)$, (4.3.2) holds.

For $\nu = 0$, since $B_2(x) = x(x-1) + \frac{1}{6}$, $B_2 = \frac{1}{6}$, there occurs again the formula of the inscribed trapezium [see (4.2.6), (4.2.7) for $a = 0$, $b = 1$].

We know (see § 3.1) that for $0 < x < 1$ the polynomial $B_{2n}(x) - B_{2n}$ $(n > 0)$ has sign opposite to B_{2n}, that is has the sign of $(-1)^n$; we also know [see (3.1.20)] that the maximum of $|B_{2n}(x) - B_{2n}|$ is obtained for $x = \frac{1}{2}$ and its value is $2\left(1 - \frac{1}{2^{2n}}\right)|B_{2n}|$. Therefore from (4.3.2) we get

$$|R(u)| \leq \frac{1}{(2\nu+2)!}\,2\left(1 - \frac{1}{2^{2\nu+2}}\right)|B_{2\nu+2}|\int_0^1 |u^{(2\nu+2)}(x)|\,dx$$

$$= 2\left(1 - \frac{1}{2^{2\nu+2}}\right)\frac{|B_{2\nu+2}|}{(2\nu+2)!}\,V_{2\nu+1} \tag{4.3.4}$$

where $V_{2\nu+1}$ denotes the *total variation* in [0, 1] of the absolutely continuous function $u^{(2\nu+1)}(x)$.

Under the hypothesis that $u^{(2\nu+2)}(x)$ *be bounded in* [0, 1], namely $M_{2\nu+2} = \sup_{0\le x\le 1} |u^{(2\nu+2)}(x)|$ and from the fact that $B_{2\nu+2}(x) - B_{2\nu+2}$ does not change its sign in [0, 1] and from $\int_0^1 B_{2\nu+2}(x)\,dx = 0$ [see (3.1.13)], we deduce

$$|R(u)| \le M_{2\nu+2} \int_0^1 \frac{|B_{2\nu+2}(x) - B_{2\nu+2}|}{(2\nu+2)!}\,dx = M_{2\nu+2} \left| \int_0^1 \frac{B_{2\nu+2}(x) - B_{2\nu+2}}{(2\nu+2)!}\,dx \right|$$

$$= M_{2\nu+2} \left| -\frac{B_{2\nu+2}}{(2\nu+2)!} \right| = \frac{|B_{2\nu+2}|}{(2\nu+2)!} M_{2\nu+2}\,, \tag{4.3.5}$$

while, if $u^{(2\nu+2)}(x)$ *is continuous in* [0, 1] the mean value theorem gives

$$R(u) = u^{(2\nu+2)}(\xi) \int_0^1 \frac{B_{2\nu+2}(x) - B_{2\nu+2}}{(2\nu+2)!}\,dx = -\frac{B_{2\nu+2}}{(2\nu+2)!} u^{(2\nu+2)}(\xi)\,,$$

$$(0 < \xi < 1)\,. \tag{4.3.6}$$

For $\nu = 0$, (4.3.4), (4.3.5), (4.3.6) reduce to (4.2.8), (4.2.9), (4.2.10) written with $a = 0$, $b = 1$.

4.4. Other examples of application of the general rule. Let us write an elementary quadrature formula on a finite interval of the type $[-a, a]$ with

$$g(x) = 1; \quad m = 3\,, \quad x_1 = -a\,, \quad x_2 = 0\,, \quad x_3 = a$$

which will be exact for the functions 1, $\cosh x$, $\sinh x$. It is clear that we have to assume $E = \frac{d^3}{dx^3} - \frac{d}{dx}$, whence the equation $E^*(\varphi) = g$ must be written $\varphi''' - \varphi' + 1 = 0$. We are not interested in its solutions $\varphi_0(x)$, $\varphi_3(x)$, but we need two arbitrary solutions $\varphi_1(x)$, $\varphi_2(x)$ for which we may take

$$\left.\begin{aligned} \varphi_1(x) &= A_1 + B_1 \cosh x + C_1 \sinh x + x\,, \\ \varphi_2(x) &= A_2 + B_2 \cosh x + C_2 \sinh x + x\,, \end{aligned}\right\} \tag{4.4.1}$$

where A_1, B_1, C_1, A_2, B_2, C_2 are arbitrary constants. Then considering that $E_0^* = 1$, $E_1^* = -\frac{d}{dx}$, $E_2^* = \frac{d^2}{dx^2} - 1$ and using (4.1.1), (4.1.2), (4.1.3), we easily obtain the following formula, *valid for every* $u(x) \in A\,C^2[-a, a]$:

$$\begin{aligned} \int_{-a}^{a} u(x)\,dx &= (-A_1 + a)\,u(-a) + (A_1 - A_2)\,u(0) + (A_2 + a)\,u(a) \\ &+ (B_1 \sinh a - C_1 \cosh a - 1)\,u'(-a) + (C_1 - C_2)\,u'(0) \\ &+ (B_2 \sinh a + C_2 \cosh a + 1)\,u'(a) \\ &+ (A_1 + B_1 \cosh a - C_1 \sinh a - a)\,u''(-a) \\ &+ (-A_1 - B_1 + A_2 + B_2)\,u''(0) \\ &+ (-A_2 - B_2 \cosh a - C_2 \sinh a - a)\,u''(a) + R(u) \end{aligned} \tag{4.4.2}$$

with

$$\left.\begin{aligned} R(u) &= \int_{-a}^{a} \Phi(x)\,[u'''(x) - u'(x)]\,dx\,, \\ \Phi(x) &= \begin{cases} A_1 + B_1 \cosh x + C_1 \sinh x + x\,, & (-a \leqq x < 0)\,, \\ A_2 + B_2 \cosh x + C_2 \sinh x + x\,, & (0 < x \leqq a)\,. \end{cases} \end{aligned}\right\} \qquad (4.4.3)$$

From (4.4.2) we may deduce infinitely many formulae by assigning the values of the 6 parameters A_1, B_1, C_1, A_2, B_2, C_2. Let us see whether it is possible to deduce one of them such that the values of the derivatives of $u(x)$ do not occur. We must therefore choose the above-mentioned parameters in a way that we may obtain

$$B_1 \sinh a - C_1 \cosh a - 1 = 0\,, \qquad C_1 - C_2 = 0\,,$$
$$B_2 \sinh a + C_2 \cosh a + 1 = 0\,,$$
$$A_1 + B_1 \cosh a - C_1 \sinh a - a = 0\,,\ -A_1 - B_1 + A_2 + B_2 = 0\,,$$
$$-A_2 - B_2 \cosh a - C_2 \sinh a - a = 0\,.$$

This system with 6 linear equations in 6 unknowns admits of one and only one solution, given by

$$A_1 = -B_1 = -A_2 = B_2 = \frac{a \cosh a - \sinh a}{\cosh a - 1}\,, \qquad C_1 = C_2 = 1 - \frac{a \sinh a}{\cosh a - 1}\,.$$

Then, putting

$$\lambda = \frac{\sinh a - a}{\cosh a - 1}\,, \qquad (4.4.4)$$

we easily find that (4.4.2), (4.4.3) can be written as

$$\boxed{\int_{-a}^{a} u(x)\,dx = \lambda\, u(-a) + 2\,(a - \lambda)\, u(0) + \lambda\, u(a) + R(u)}\,, \qquad (4.4.5)$$

$$\left.\begin{aligned} R(u) &= \int_{-a}^{a} \Phi(x)\,[u'''(x) - u'(x)]\,dx\,, \\ \Phi(x) &= \begin{cases} \varphi_1(x) = \lambda\,[\cosh (x + a) - 1] - [\sinh (x + a) - (x + a)]\,, \\ \qquad (-a \leqq x \leqq 0)\,, \\ \varphi_2(x) = -\lambda\,[\cosh (x - a) - 1] - [\sinh (x - a) - (x - a)]\,, \\ \qquad (0 \leqq x \leqq a)\,. \end{cases} \end{aligned}\right\} \qquad (4.4.6)$$

The influence-function $\Phi(x)$ is symmetric with respect to the origin [since $\varphi_1(x) + \varphi_2(-x) = 0$] and is as in Fig. 4. [1]) Therefore

$$|R(u)| \leqq \left(\log \frac{1 + \lambda}{1 - \lambda} - 2\,\lambda\right) \int_{-a}^{a} \left|\frac{d}{dx}\,[u''(x) - u(x)]\right| dx = \left(\log \frac{1 + \lambda}{1 - \lambda} - 2\,\lambda\right) V\,. \qquad (4.4.7)$$

[1]) From (4.4.4) we may easily deduce $0 < \lambda < 1$, $2\,\lambda < \log \frac{1 + \lambda}{1 - \lambda} < a$. Note that $\Phi(x)$ is of class C^1 in $[-a, a]$.

follows from (4.4.6), where V is the *total variation* of the absolutely continuous function $u''(x) - u(x)$.

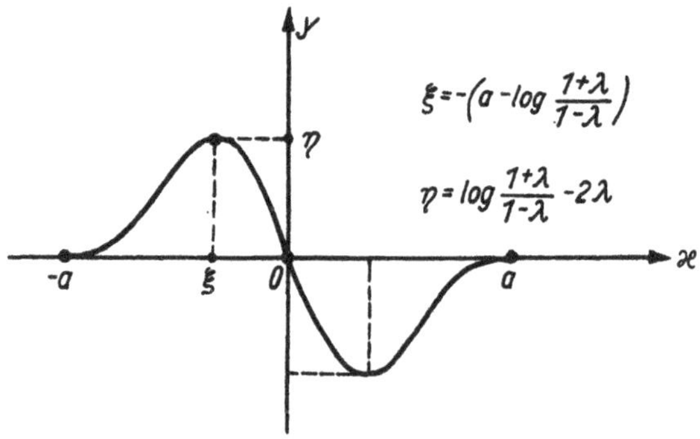

Fig. 4

Moreover if we assume $u'''(x)$ [and therefore $u'''(x) - u'(x)$] *bounded in* $[-a, a]$ and put $M = \sup\limits_{-a \leq x \leq a} |u'''(x) - u'(x)|$ we may write

$$|R(u)| \leqq M \int_{-a}^{a} |\Phi(x)|\, dx = 2\, M \int_{-a}^{0} \varphi_1(x)\, dx = M \left[2\, \frac{\frac{a}{2} \cosh \frac{a}{2} - \sinh \frac{a}{2}}{\sinh \frac{a}{2}} \right]^2 . \tag{4.4.8}$$

Similarly, *assuming* $0 < a < 2\pi$, we may obtain an elementary quadrature formula on the interval $[-a, a]$ with

$$g(x) = 1; \quad m = 3, \quad x_1 = -a, \quad x_2 = 0, \quad x_3 = a$$

which is exact for the functions 1, $\cos x$, $\sin x$. Assuming $E = \frac{d^3}{dx^3} + \frac{d}{dx}$ and *making the hypothesis* $u(x) \in A\ C^2[-a, a]$ we get the following formulae (where 6 arbitrary parameters A_1, B_1, C_1, A_2, B_2, C_2 occur):

$$\begin{aligned} \int_{-a}^{a} u(x)\, dx = {} & (A_1 + a)\, u(-a) + (A_2 - A_1)\, u(0) + (-A_2 + a)\, u(a) \\ & + (-B_1 \sin a - C_1 \cos a + 1)\, u'(-a) + (C_1 - C_2)\, u'(0) \\ & + (-B_2 \sin a + C_2 \cos a - 1)\, u'(a) \\ & + (A_1 + B_1 \cos a - C_1 \sin a + a)\, u''(-a) \\ & + (A_2 - A_1 + B_2 - B_1)\, u''(0) \\ & + (-A_2 - B_2 \cos a - C_2 \sin a + a)\, u''(a) + R(u), \end{aligned} \tag{4.4.9}$$

$$\left.\begin{aligned} R(u) &= \int_{-a}^{a} \Phi(x)\,[u'''(x) + u'(x)]\,dx\,, \\ \Phi(x) &= \begin{cases} \varphi_1(x) = A_1 + B_1 \cos x + C_1 \sin x - x\,, & (-a \leqq x < 0)\,, \\ \varphi_2(x) = A_2 + B_2 \cos x + C_2 \sin x - x\,, & (0 < x \leqq a)\,. \end{cases} \end{aligned}\right\} \quad (4.4.10)$$

Assuming

$$-A_1 = B_1 = A_2 = -B_2 = \frac{\sin a - a \cos a}{1 - \cos a}\,, \qquad C_1 = C_2 = \frac{a \sin a}{1 - \cos a} - 1$$

and putting

$$\lambda = \frac{a - \sin a}{1 - \cos a} > 0\,, \qquad (4.4.11)$$

(4.4.9) and (4.4.10) become

$$\boxed{\int_{-a}^{a} u(x)\,dx = \lambda\, u(-a) + 2\,(a - \lambda)\, u(0) + \lambda\, u(a) + R(u)}\,, \qquad (4.4.12)$$

$$\left.\begin{aligned} R(u) &= \int_{-a}^{a} \Phi(x)\,[u'''(x) + u'(x)]\,dx\,, \\ \Phi(x) &= \begin{cases} \varphi_1(x) = \lambda\,[1 - \cos(x + a)] - [(x + a) - \sin(x + a)]\,, \\ \qquad (-a \leqq x \leqq 0)\,, \\ \varphi_2(x) = -\lambda\,[1 - \cos(x - a)] - [(x - a) - \sin(x - a)]\,, \\ \qquad (0 \leqq x \leqq a)\,. \end{cases} \end{aligned}\right\} \quad (4.4.13)$$

The influence-function $\Phi(x)$ has a graph analogous to that of Fig. 4, with $\xi = -a + 2 \arctan \lambda$, $\eta = 2\,(\lambda - \arctan \lambda)$. [1]) Then it is easy to deduce from (4.4.13) the bound

$$|R(u)| \leqq 2\,(\lambda - \arctan \lambda)\, V\,, \qquad (4.4.14)$$

where V denotes the *total variation* of the absolutely continuous function $u''(x) + u(x)$. If then we assume $u'''(x)$ *bounded in* $[-a, a]$ we may write, using (4.4.11)

$$|R(u)| \leqq M \left[2\, \frac{\sin \frac{a}{2} - \frac{a}{2} \cos \frac{a}{2}}{\sin \frac{a}{2}} \right]^2, \qquad (4.4.15)$$

where $M = \sup\limits_{-a \leqq x \leqq a} |u'''(x) + u'(x)|$.

We give here another example, deriving an elementary quadrature formula on a finite interval $[a, b]$ with

$$g(x) = \frac{1}{\sqrt{x - a}}\,; \quad m = 2\,, \quad x_1 = a\,, \quad x_2 = b; \quad n = 2\,, \quad E = \frac{d^2}{dx^2}\,.$$

[1]) It is understood that $0 < \arctan \lambda < \frac{\pi}{2}$. It is easy to see that we have $\arctan \lambda < \frac{a}{2}$ (this is evident if $\pi \leqq a < 2\pi$).

The differential equation $E^*(\varphi) = g$ is written $\varphi'' = \frac{1}{\sqrt{x-a}}$; we do not need its solution $\varphi_0(x)$, $\varphi_2(x)$ and it suffices to fix an arbitrary solution $\varphi_1(x)$ which will be of the type

$$\varphi_1(x) = A\,x + B + \frac{4}{3}(x-a)^{3/2}\,, \tag{4.4.16}$$

where A and B are arbitrary constants. Then an immediate application of (4.1.1), (4.1.2) and (4.1.3), leads, *under the hypothesis* $u(x) \in A\,C^1[a, b]$, to the following formulae

$$\int_a^b \frac{u(x)\,dx}{\sqrt{x-a}} = -A\,u(a) + [A + 2\,(b-a)^{1/2}]\,u(b)$$

$$+ (A\,a + B)\,u'(a) - \left[A\,b + B + \frac{4}{3}(b-a)^{3/2}\right]u'(b) + R(u)\,, \tag{4.4.17}$$

$$R(u) = \int_a^b \varphi_1(x)\,u''(x)\,dx\,. \tag{4.4.18}$$

In (4.4.17) we may drop the terms with the first derivatives if

$$A = -\frac{4}{3}(b-a)^{1/2}\,, \quad B = \frac{4}{3}a\,(b-a)^{1/2};$$

thus we obtain

$$\boxed{\int_a^b \frac{u(x)\,dx}{\sqrt{x-a}} = \frac{2}{3}\sqrt{b-a}\,[2\,u(a) + u(b)] + R(u)}\,, \tag{4.4.19}$$

$$R(u) = \int_a^b \Phi(x)\,u''(x)\,dx\,,\quad \Phi(x) = \varphi_1(x) = -\frac{4}{3}(x-a)\left(\sqrt{b-a} - \sqrt{x-a}\right). \tag{4.4.20}$$

In $[a, b]$ we have $\Phi(x) \leqq 0$. The maximum of $|\Phi(x)|$ is attained for $x = a + \frac{4}{9}(b-a)$ and has the value $\frac{16}{81}(b-a)^{3/2}$; therefore

$$|R(u)| \leqq \frac{16}{81}(b-a)^{3/2}\,V_1 \tag{4.4.21}$$

follows from (4.4.20), where V_1 is the *total variation* of the absolutely continuous function $u'(x)$.

If we assume $u''(x)$ *bounded in* $[a, b]$, from (4.4.20) we may deduce the bound:

$$|R(u)| \leqq M_2 \int_a^b |\Phi(x)|\,dx = \frac{2}{15}(b-a)^{5/2}\,M_2\,, \tag{4.4.22}$$

where $M_2 = \sup\limits_{a \leq x \leq b} |u''(x)|$. Moreover if $u''(x)$ *is continuous* in $[a, b]$ we may also write

$$R(u) = -\frac{2}{15}(b-a)^{5/2}\, u''(\xi)\,, \quad (a < \xi < b)\,. \tag{4.4.23}$$

Let us give a fourth example, assuming

$$a = 1\,, \quad b = +\infty; \quad g(x) = \frac{1}{x^\alpha} \quad (\alpha > 0);$$

$$m = 2\,, \quad x_1 = 1\,, \quad x_2 = 1 + h \quad (h > 0);$$

$$n = 2\,, \quad E = \frac{d^2}{dx^2}\,,$$

and adopting for the weight function $g(x)$ (that is for the exponent α) and for the argument function $u(x)$, hypotheses (2.2.11) and (2.2.12).

We therefore recall that the equation $E(u) = 0$ has the two solutions $u_1 = 1$, $u_2 = x$ and that the associated solutions of $E^*(v) = 0$, according to (1.3.5), are $v_1 = -x$, $v_2 = 1$. Then formula (2.2.11) requires that $\frac{1}{x^\alpha} \cdot 1$, $\frac{1}{x^\alpha} \cdot x \in L[1, +\infty)$ and these are satisfied if and only if $\alpha > 2$, so that we can put $\alpha = 2 + \sigma$ (with $\sigma > 0$). Equation (2.2.12) then requires, in order that

$$-x\, u''(x) \int\limits_x^{+\infty} 1 \cdot \frac{1}{\xi^{2+\sigma}}\, d\xi\,, \quad 1 \cdot u''(x) \int\limits_x^{+\infty} \xi\, \frac{1}{\xi^{2+\sigma}}\, d\xi \in L[1, +\infty)$$

hold, that they reduce to

$$\frac{u''(x)}{x^\sigma} \in L[1, +\infty)\,. \tag{4.4.24}$$

We can obtain through this hypothesis the wanted quadrature formula for the integral (absolutely convergent) $\int\limits_1^{+\infty} \frac{u(x)}{x^{2+\sigma}}\, dx$, under the condition that it be exact when $u(x)$ is a polynomial of first degree. The differential equation $E^*(\varphi) = g$ is written $\varphi'' = \frac{1}{x^{2+\sigma}}$ and has the general solution $\varphi(x) = A\, x + B + \frac{1}{\sigma(1+\sigma)} \frac{1}{x^\sigma}$ (with A and B as arbitrary constants). Of the two solutions $\varphi_0(x)$, $\varphi_2(x)$ we are interested only in the second one, for which we obviously have

$$\varphi_2(x) = \frac{1}{\sigma(1+\sigma)} \frac{1}{x^\sigma}\,,$$

while we must choose arbitrarily another solution

$$\varphi_1(x) = A\, x + B + \frac{1}{\sigma(1+\sigma)} \frac{1}{x^\sigma}\,.$$

Next an immediate application of (4.1.1), (4.1.2) and (4.1.3) leads to the following quadrature formula, *valid under the hypothesis* (4.4.24) and depending on the two arbitrary parameters A and B:

$$\int_1^{+\infty} \frac{u(x)\,dx}{x^{2+\sigma}} = \left(-A + \frac{1}{1+\sigma}\right) u(1) + A\,u\,(1+h) + \left(A + B + \frac{1}{\sigma(1+\sigma)}\right)$$
$$\times\, u'(1) - [A\,(1+h) + B]\,u'\,(1+h) + R(u)\,, \qquad (4.4.25)$$

with

$$R(u) = \int_1^{1+h} (A\,x + B)\,u''(x)\,dx + \frac{1}{\sigma(1+\sigma)} \int_1^{+\infty} \frac{u''(x)}{x^{\sigma}}\,dx\,. \qquad (4.4.26)$$

We may choose the values of the parameters A and B in such a way as to eliminate the values of the derivatives; it suffices to put

$$A = \frac{1}{\sigma(1+\sigma)}\,\frac{1}{h}\,, \qquad B = -\frac{1}{\sigma(1+\sigma)}\left(1 + \frac{1}{h}\right),$$

which transforms (4.4.25) and (4.4.26) into the following formulae:

$$\boxed{\int_1^{+\infty} \frac{u(x)\,dx}{x^{2+\sigma}} = \frac{1}{\sigma(1+\sigma)}\left[\left(\sigma - \frac{1}{h}\right) u(1) + \frac{1}{h}\,u\,(1+h)\right] + R(u)}\,, \qquad (4.4.27)$$

$$R(u) = \frac{1}{\sigma(1+\sigma)}\left\{\int_1^{1+h}\left(\frac{x-1}{h} - 1\right) u''(x)\,dx + \int_1^{+\infty} \frac{u''(x)}{x^{\sigma}}\,dx\right\}. \qquad (4.4.28)$$

From (4.4.28) we may immediately deduce a bound for $|R(u)|$ in the case where, for instance, there holds a bound on $u''(x)$ of the type $|u''(x)| \leqq \frac{M}{x^{1+\tau-\sigma}}$ ($\tau > 0$), which implies that (4.4.24) holds. We leave to the reader the easy computation.

4.5. The Gauss-Jacobi formulae. The Gauss-Jacobi elementary quadrature formulae are relative to the interval $[-1, 1]$ and to the weight $g(x) = (1-x)^{\alpha} \times (1+x)^{\beta}$, with $\alpha > -1$, $\beta > -1$ so that $g(x) \in L[-1, 1]$. More precisely they are of the type

$$\int_{-1}^{1} (1-x)^{\alpha}\,(1+x)^{\beta}\,u(x)\,dx = \sum_{i=1}^{m} A_i\,u(x_i) + R(u) \qquad (4.5.1)$$

with m nodes ($m = 1, 2, 3, \ldots$), under the condition that $R(u) = 0$ *when* $u(x)$ *is a polynomial of degree* $\leqq 2\,m - 1$.

Remembering what we said in § 2.5 we see that working out (4.5.1) is equivalent to a *Gauss problem* with $n = 2\,m$, $E = \frac{d^{2m}}{dx^{2m}}$, $p = n - 1 = 2\,m - 1$.

Therefore we must consider the following homogeneous differential boundary problem [see (2.5.6)]:

$$\frac{d^{2m}u}{dx^{2m}} = 0\,, \qquad u(x_1) = u(x_2) = \cdots = u(x_m) = 0$$

which has the following m linearly independent solutions

$$U_r(x) = x^r \prod_{i=1}^{m} (x - x_i)\,, \qquad (r = 0, 1, \ldots, m-1)\,.$$

It follows, according to theor. 2.5.I, that (4.5.1) holds if and only if the nodes are chosen so as to satisfy

$$\int_{-1}^{1} (1-x)^\alpha (1+x)^\beta x^r \prod_{i=1}^{m} (x - x_i)\, dx = 0\,, \qquad (r = 0, 1, \ldots, m-1)\,.$$

From § 3.4 this can happen if and only if the polynomial $\prod_{i=1}^{m} (x - x_i)$ coincides, apart from a constant factor, with the Jacobi polynomial $P_m^{(\alpha,\beta)}(x)$, that is if and only if *the nodes* $x_1, x_2, \ldots, x_m$ *are the zeros* $x_{m1}^{(\alpha,\beta)} < x_{m2}^{(\alpha,\beta)} < \cdots < x_{mm}^{(\alpha,\beta)}$ *of the Jacobi polynomials* $P_m^{(\alpha,\beta)}(x)$.

Choosing the nodes in this way, a formula of the type (4.5.1) can therefore exist, and from theor. 2.5.I, it is even necessarily unique (since we have $m(n-p) - n + q = m \cdot 1 - 2m + m = 0$). For the validity of (4.5.1) *we must then set the condition* $u(x) \in A\ C^{2m-1}[-1, 1]$.

Writing again (4.5.1) as

$$\boxed{\int_{-1}^{1} (1-x)^\alpha (1+x)^\beta u(x)\, dx = \sum_{i=1}^{m} H_{mi}^{(\alpha,\beta)} u(x_{mi}^{(\alpha,\beta)}) + R(u)}\,, \qquad (4.5.2)$$

let us now compute the coefficients $H_{mi}^{(\alpha,\beta)}$ (which are called *Christoffel numbers*) and then the remainder $R(u)$, following the scheme given in § 2.4.

Let us put in (4.5.2)

$$u(x) = \left[\frac{d}{dx} P_m^{(\alpha,\beta)}(x)\right]^{-1}_{x = x_{mj}^{(\alpha,\beta)}} \cdot \frac{P_m^{(\alpha,\beta)}(x)}{x - x_{mj}^{(\alpha,\beta)}}\,, \qquad (j = 1, 2, \ldots, m)\,,$$

bearing in mind that this is a polynomial of degree $m-1$ [whence we have $R(u) = 0$] and that we evidently have $u(x_{mi}^{(\alpha,\beta)}) = \delta_{ij}$; then we immediately find

$$H_{mj}^{(\alpha,\beta)} = \left[\frac{d}{dx} P_m^{(\alpha,\beta)}(x)\right]^{-1}_{x = x_{mj}^{(\alpha,\beta)}} \cdot \int_{-1}^{1} (1-x)^\alpha (1+x)^\beta \frac{P_m^{(\alpha,\beta)}(x)}{x - x_{mj}^{(\alpha,\beta)}}\, dx\,,$$

$$(j = 1, 2, \ldots, m)\,. \qquad (4.5.3)$$

The integral occurring here can be easily computed.

Let us see again the Christoffel-Darboux summation formula [see (3.4.15)] and put in it $y = x_{mj}^{(\alpha,\beta)}$ thus obtaining

$$\sum_{k=0}^{m-1} \frac{P_k^{(\alpha,\beta)}(x)\, P_k^{(\alpha,\beta)}(x_{mj}^{(\alpha,\beta)})}{h_k^{(\alpha,\beta)}} = -\frac{A_m^{(\alpha,\beta)}}{h_m^{(\alpha,\beta)}} \frac{P_m^{(\alpha,\beta)}(x)\, P_{m+1}^{(\alpha,\beta)}(x_{mj}^{(\alpha,\beta)})}{x - x_{mj}^{(\alpha,\beta)}}\,.$$

Let us now multiply both members of this relation by $(1-x)^\alpha (1+x)^\beta$ and integrate over the interval $[-1, 1]$; bearing in mind the properties of orthogonality of the Jacobi polynomials, we thus obtain

$$\frac{1}{h_0^{(\alpha,\beta)}} \int_{-1}^{1} (1-x)^\alpha (1+x)^\beta \, dx = - \frac{A_m^{(\alpha,\beta)} P_{m+1}^{(\alpha,\beta)}(x_{mj}^{(\alpha,\beta)})}{h_m^{(\alpha,\beta)}} \times \int_{-1}^{1} (1-x)^\alpha (1+x)^\beta \frac{P_m^{(\alpha,\beta)}(x)}{x - x_{mj}^{(\alpha,\beta)}} \, dx ,$$

from which, remembering that $\int_{-1}^{1} (1-x)^\alpha (1+x)^\beta \, dx = h_0^{(\alpha,\beta)}$ [see (3.4.12)] we get

$$\int_{-1}^{1} (1-x)^\alpha (1+x)^\beta \frac{P_m^{(\alpha,\beta)}(x)}{x - x_{mj}^{(\alpha,\beta)}} \, dx = - \frac{h_m^{(\alpha,\beta)}}{A_m^{(\alpha,\beta)} P_{m+1}^{(\alpha,\beta)}(x_{mj}^{(\alpha,\beta)})} . \tag{4.5.4}$$

We may also express $P_{m+1}^{(\alpha,\beta)}(x_{mj}^{(\alpha,\beta)})$, considering (3.4.30) and putting in it $x = x_{mj}^{(\alpha,\beta)}$; this gives

$$P_{m+1}^{(\alpha,\beta)}(x_{mj}^{(\alpha,\beta)}) = - \frac{(\alpha+\beta+2m+2)\,[1-(x_{mj}^{(\alpha,\beta)})^2]}{2(m+1)(\alpha+\beta+m+1)} \left[\frac{d}{dx} P_m^{(\alpha,\beta)}(x)\right]_{x=x_{mj}^{(\alpha,\beta)}} . \tag{4.5.5}$$

Through (4.5.3), (4.5.4), (4.5.5) and bearing also in mind the expressions of $h_m^{(\alpha,\beta)}$, $A_m^{(\alpha,\beta)}$ given by (3.4.13), (3.4.16) we finally obtain

$$H_{mj}^{(\alpha,\beta)} = \frac{2^{\alpha+\beta+1}\,\Gamma(\alpha+m+1)\,\Gamma(\beta+m+1)}{m!\,\Gamma(\alpha+\beta+m+1)\,[1-(x_{mj}^{(\alpha,\beta)})^2]} \cdot \left[\frac{d}{dx} P_m^{(\alpha,\beta)}(x)\right]^{-2}_{x=x_{mj}^{(\alpha,\beta)}} , \quad (j = 1, 2, \ldots, m) , \tag{4.5.6}$$

which shows, among other things, that the *Christoffel numbers are all positive*.

To obtain the expression for the remainder $R(u)$ according to (4.1.3) we must obtain the solutions $\varphi_0(x), \varphi_1(x), \ldots, \varphi_m(x)$ of the differential equation $E^*(\varphi) = g$ which is now written $\varphi^{(2m)} = (1-x)^\alpha (1+x)^\beta$. Having already computed the coefficients $H_{mi}^{(\alpha,\beta)}$ of the quadrature formula, we may apply theor. 2.4.I which, since $K(x, \xi) = \frac{(x-\xi)^{2m-1}}{(2m-1)!}$, gives

$$\varphi_i(x) = \int_{-1}^{x} \frac{(x-\xi)^{2m-1}}{(2m-1)!} (1-\xi)^\alpha (1+\xi)^\beta \, d\xi - \sum_{j=1}^{i} H_{mj}^{(\alpha,\beta)} \frac{(x - x_{mj}^{(\alpha,\beta)})^{2m-1}}{(2m-1)!} , \quad (i = 0, 1, \ldots, m) .\,[1] \tag{4.5.7}$$

[1]) We remember [see 2.4.6] that for $i = m$ we may also write

$$\varphi_m(x) = - \int_{x}^{1} \frac{(x-\xi)^{2m-1}}{(2m-1)!} (1-\xi)^\alpha (1+\xi)^\beta \, d\xi .$$

We may therefore assert

$$R(u) = \int_{-1}^{1} \Phi(x)\, u^{(2m)}(x)\, dx = \sum_{i=0}^{m} \int_{x_{mi}^{(\alpha,\beta)}}^{x_{m,i+1}^{(\alpha,\beta)}} \varphi_i(x)\, u^{(2m)}(x)\, dx\,, \quad (x_{m0}^{(\alpha,\beta)} = -1\,,\ x_{m,m+1}^{(\alpha,\beta)} = 1)\,, \tag{4.5.8}$$

with $\varphi_i(x)$ expressed by (4.5.7) and the influence-function $\Phi(x)$ by

$$\Phi(x) = \varphi_i(x) \quad \text{for} \quad x_{mi}^{(\alpha,\beta)} < x < x_{m,i+1}^{(\alpha,\beta)}\,, \quad (i = 0, 1, \ldots, m)\,. \tag{4.5.9}$$

In order to obtain from (4.5.8) some formulae of bounding and evaluation of the remainder, let us begin by stating two theorems.

Theorem 4.5.I. *The influence-function $\Phi(x)$ defined by (4.5.9) [together with (4.5.7)] belongs to the class $C^{2m-2}[-1, 1]$ and is positive inside the interval.*

Proof. From (4.5.7) and (4.5.9) it follows, differentiating k times (with $0 \leqq k \leqq 2m - 1$):

$$\Phi^{(k)}(x) = \varphi_i^{(k)}(x) \quad \text{for} \quad x_{mi}^{(\alpha,\beta)} < x < x_{m,i+1}^{(\alpha,\beta)}\,, \qquad (i = 0, 1, \ldots, m)\,, \tag{4.5.10}$$

$$\varphi_i^{(k)}(x) = \int_{-1}^{x} \frac{(x-\xi)^{2m-k-1}}{(2m-k-1)!} (1-\xi)^\alpha (1+\xi)^\beta\, d\xi - \sum_{j=1}^{i} H_{mj}^{(\alpha,\beta)} \frac{(x - x_{mj}^{(\alpha,\beta)})^{2m-k-1}}{(2m-k-1)!}\,.^{1)} \tag{4.5.11}$$

From these formulae we soon see that:

$$\Phi^{(k)}(-1) = \Phi^{(k)}(1) = 0\,, \tag{4.5.12}$$

$$\Phi^{(k)}(x) > 0 \quad \text{for} \quad -1 < x < x_{m1}^{(\alpha,\beta)}\,, \quad (-1)^k\, \Phi^{(k)}(x) > 0$$
$$\text{for} \quad x_{mm}^{(\alpha,\beta)} < x < 1\,, \quad (k = 0, 1, \ldots, 2m - 1)\,, \tag{4.5.13}$$

$$\Phi(x),\ \Phi'(x), \ldots, \Phi^{(2m-2)}(x) \text{ continuous in } [-1, 1]\,, \tag{4.5.14}$$

while $\Phi^{(2m-1)}(x)$ has discontinuities of first kind at the points $x_{m1}^{(\alpha,\beta)}, x_{m2}^{(\alpha,\beta)}, \ldots, x_{mm}^{(\alpha,\beta)}$.

The function $\Phi^{(2m-2)}(x)$ vanishes at most twice in each interval $[x_{mi}^{(\alpha,\beta)}, x_{m,i+1}^{(\alpha,\beta)}]$, $(i = 1, 2, \ldots, m - 1)$. In fact, if it should vanish 3 times, on the ground of the Rolle theorem, $\Phi^{(2m-1)}(x)$ would vanish at least twice inside $[x_{mi}^{(\alpha,\beta)}, x_{m,i+1}^{(\alpha,\beta)}]$ and this is absurd since, in $(x_{mi}^{(\alpha,\beta)}, x_{m,i+1}^{(\alpha,\beta)})$, $\Phi^{(2m-1)}(x)$ coincides with the *increasing* function $\varphi_i^{(2m-1)}(x) = \int_{-1}^{x} (1-\xi)^\alpha (1+\xi)^\beta\, d\xi - \sum_{j=1}^{i} H_{mj}^{(\alpha,\beta)}$.
From this statement and from (4.5.13) there obviously follows that $\Phi^{(2m-2)}(x)$ *vanishes at most $2m - 2$ times inside* $[-1, 1]$. We can now show that $\Phi(x)$ does not vanish inside $[-1, 1]$ and therefore is here positive, by virtue of (4.5.13). In fact, if $\Phi(x)$ should vanish at a point internal in $[-1, 1]$, using

1) For $i = m$ we may also write $\varphi_m^{(k)}(x) = -\int_{x}^{1} \frac{(x-\xi)^{2m-k-1}}{(2m-k-1)!} (1-\xi)^\alpha (1+\xi)^\beta\, d\xi$.

(4.3.12) and the Rolle theorem, we should have that, inside $[-1, 1]$, $\Phi'(x)$ would vanish at least two times, $\Phi''(x)$ at least three times, ..., $\Phi^{(2m-2)}(x)$ at least $2m - 1$ times, which is in contraposition with the preceding statement. The theorem is thus proved.

Theorem 4.5.II. *We have*

$$\int_{-1}^{1} \Phi(x)\, dx = \frac{2^{\alpha+\beta+2m+1}\, m!\, \Gamma(\alpha+m+1)\, \Gamma(\beta+m+1)\, \Gamma(\alpha+\beta+m+1)}{(2m)!\, (\alpha+\beta+2m+1)\, [\Gamma(\alpha+\beta+2m+1)]^2}. \tag{4.5.15}$$

Proof. Let us examine again (4.5.2) and put in it $u(x) = [P_m^{(\alpha,\beta)}(x)]^2 = [a_m x^m + \cdots]^2$. Considering also (4.5.8) we obtain

$$\int_{-1}^{1} (1-x)^\alpha (1+x)^\beta [P_m^{(\alpha,\beta)}(x)]^2\, dx = (2m)!\, a_m^2 \int_{-1}^{1} \Phi(x)\, dx.$$

The value of the integral in the left member is $h_m^{(\alpha,\beta)}$ [see (3.4.12)]; from it, remembering the values of $h_m^{(\alpha,\beta)}$ and a_m [see (3.4.13), (3.4.8)] we immediately obtain (4.5.15).

Let us now consider again the expression (4.5.8) of the remainder $R(u)$. We immediately obtain, with only the hypothesis $u(x) \in A\, C^{2m-1}[-1, 1]$:

$$|R(u)| \leqq \max_{-1 \leqq x \leqq 1} \Phi(x) \cdot V_{2m-1}, \tag{4.5.16}$$

where V_{2m-1} denotes the *total variation* of the absolutely continuous function $u^{(2m-1)}(x)$. From the reasoning of the proof of theorem 4.5.I we see that, if $m > 1$, inside $[-1, 1]$, $\Phi'(x)$ vanishes at one and only one point x_0; therefore

$$\max_{-1 \leqq x \leqq 1} \Phi(x) = \Phi(x_0), \tag{4.5.17}$$

but it does not seem easy to evaluate the general expression of $\Phi(x_0)$ in terms of m, α, β. Therefore (4.5.16) can be used at most for the first values of m.

If on the contrary we accept the hypothesis that $u^{(2m)}(x)$ *be bounded* in $[-1, 1]$, and put $M_{2m} = \sup_{-1 \leqq x \leqq 1} |u^{(2m)}(x)|$ we may deduce from (4.5.8) by virtue of theorems 4.5.I and 4.5.II:

$$|R(u)| \leqq \frac{2^{\alpha+\beta+2m+1}\, m!\, \Gamma(\alpha+m+1)\, \Gamma(\beta+m+1)\, \Gamma(\alpha+\beta+m+1)}{(2m)!\, (\alpha+\beta+2m+1)\, [\Gamma(\alpha+\beta+2m+1)]^2} M_{2m}. \tag{4.5.18}$$

Under the more restrictive hypothesis that $u^{(2m)}(x)$ *be continuous in* $[-1, 1]$ we may write, remembering that $\Phi(x) \geqq 0$ and applying the mean value theorem:

$$R(u) = \frac{2^{\alpha+\beta+2m+1}\, m!\, \Gamma(\alpha+m+1)\, \Gamma(\beta+m+1)\, \Gamma(\alpha+\beta+m+1)}{(2m)!\, (\alpha+\beta+2m+1)\, [\Gamma(\alpha+\beta+2m+1)]^2} u^{(2m)}(\xi), \quad (-1 < \xi < 1). \tag{4.5.19}$$

4.6. Particular cases of the Gauss-Jacobi formulae. The first particular case which we wish to examine is that in which

$$\alpha = \beta = \lambda - \frac{1}{2} \quad \text{with} \quad \lambda > -\frac{1}{2}, \quad \lambda \neq 0$$ [1])

whence (4.5.2) is to be written

$$\boxed{\int_{-1}^{1} (1 - x^2)^{\lambda - \frac{1}{2}} u(x)\, dx = \sum_{i=1}^{m} H_{mi}^{(\lambda)} u(x_{mi}^{(\lambda)}) + R(u)}, \tag{4.6.1}$$

having put

$$H_{mi}^{(\lambda)} = H_{mi}^{\left(\lambda - \frac{1}{2}, \lambda - \frac{1}{2}\right)}, \quad x_{mi}^{(\lambda)} = x_{mi}^{\left(\lambda - \frac{1}{2}, \lambda - \frac{1}{2}\right)}. \tag{4.6.2}$$

In this case, instead of the Jacobi polynomials $P_m^{\left(\lambda - \frac{1}{2}, \lambda - \frac{1}{2}\right)}(x)$, we may consider the *ultraspherical polynomials* $P_m^{(\lambda)}(x)$ defined by [see (3.5.2)]:

$$P_m^{(\lambda)}(x) = \frac{\Gamma\left(\lambda + \frac{1}{2}\right) \Gamma(2\lambda + m)}{\Gamma(2\lambda)\, \Gamma\left(\lambda + m + \frac{1}{2}\right)} P_m^{\left(\lambda - \frac{1}{2}, \lambda - \frac{1}{2}\right)}(x). \tag{4.6.3}$$

These polynomials are even or odd functions according to whether m is even or odd. Therefore the nodes $x_{mi}^{(\lambda)}$, zeros of $P_m^{(\lambda)}(x)$, lie symmetrically with respect to the point $x = 0$; that is we have

$$x_{mi}^{(\lambda)} = -x_{m, m-i+1}^{(\lambda)}. \tag{4.6.4}$$

Putting $\alpha = \beta = \lambda - \frac{1}{2}$ in (4.5.6) and using (4.6.3), we obtain

$$H_{mj}^{(\lambda)} = \frac{2^{2\lambda}\, \Gamma(2\lambda + m) \left[\Gamma\left(\lambda + \frac{1}{2}\right)\right]^2}{m!\, [\Gamma(2\lambda)]^2 \left[1 - (x_{mj}^{(\lambda)})^2\right]} \left[\frac{d}{dx} P_m^{(\lambda)}(x)\right]_{x = x_{mj}^{(\lambda)}}^{-2};$$

this formula can be simplified using the well known formula [see (3.2.11)]:

$$\Gamma(2\lambda) = 2^{2\lambda - 1} \pi^{-\frac{1}{2}} \Gamma(\lambda)\, \Gamma\left(\lambda + \frac{1}{2}\right) \tag{4.6.5}$$

in order to express $\Gamma\left(\lambda + \frac{1}{2}\right)$. Thus we find

$$H_{mj}^{(\lambda)} = \frac{2^{2(1-\lambda)}\, \pi\, \Gamma(2\lambda + m)}{m!\, [\Gamma(\lambda)]^2 \left[1 - (x_{mj}^{(\lambda)})^2\right]} \left[\frac{d}{dx} P_m^{(\lambda)}(x)\right]_{x = x_{mj}^{(\lambda)}}^{-2} \tag{4.6.6}$$

and it is evident that for (4.6.4) there results

$$H_{mj}^{(\lambda)} = H_{m, m-j+1}^{(\lambda)}. \tag{4.6.7}$$

Bearing in mind (4.6.4) and (4.6.7), from (4.5.7) $\left(\text{with } \alpha = \beta = \lambda - \frac{1}{2}\right)$

[1]) The case $\lambda = 0$ $\left(\text{that is } \alpha = \beta = -\frac{1}{2}\right)$, excluded here, will be studied in the following: see (4.6.25).

we easily obtain $\varphi_{m-i}(x) = \varphi_i(-x)$ and therefore from (4.5.9) it follows that $\Phi(x)$ *is an even function.*

The formula (4.5.15) becomes

$$\int_{-1}^{1} \Phi(x)\,dx = \frac{2^{2\lambda+2m}\,m!\left[\Gamma\left(\lambda+m+\frac{1}{2}\right)\right]^2 \Gamma(2\lambda+m)}{(2m)!\,(2\lambda+2m)\,[\Gamma(2\lambda+2m)]^2},$$

or, using (4.6.5) to express $\Gamma(2\lambda+2m)$:

$$\int_{-1}^{1} \Phi(x)\,dx = \frac{\pi\,m!\,\Gamma(2\lambda+m)}{2^{2\lambda+2m-1}(2m)!\,(\lambda+m)\,[\Gamma(\lambda+m)]^2}. \tag{4.6.8}$$

Considering (4.5.16) and (4.5.17), from the fact that $\Phi'(x)$ is an odd function, it is evident that $x_0 = 0$; whence we have

$$|R(u)| \leqq \Phi(0)\,V_{2m-1}. \tag{4.6.9}$$

Finally, from (4.6.8), the two equations (4.5.18) and (4.5.19) become

$$|R(u)| \leqq \frac{\pi\,m!\,\Gamma(2\lambda+m)}{2^{2\lambda+2m-1}(2m)!\,(\lambda+m)\,[\Gamma(\lambda+m)]^2} M_{2m}, \tag{4.6.10}$$

$$R(u) = \frac{\pi\,m!\,\Gamma(2\lambda+m)}{2^{2\lambda+2m-1}(2m)!\,(\lambda+m)\,[\Gamma(\lambda+m)]^2} u^{(2m)}(\xi), \quad (-1<\xi<1). \tag{4.6.11}$$

The subcase

$$\lambda = \frac{1}{2} \quad (\text{that is } \alpha = \beta = 0)$$

in which the ultraspherical polynomials $P_m^{(\lambda)}(x)$ reduce to the *Legendre polynomials* $P_m(x)$ [see (3.5.5)] is of particular importance. Equation (4.6.1) then becomes the classical *Gauss quadrature formula*

$$\boxed{\int_{-1}^{1} u(x)\,dx = \sum_{i=1}^{m} H_{mi}\,u(x_{mi}) + R(u)}, \tag{4.6.12}$$

where $H_{mi} = H_{mi}^{(1/2)}$ and the nodes $x_{mi} = x_{mi}^{(1/2)}$ are the zeros of the Legendre polynomial $P_m(x)$.

Equation (4.6.6) reduces to

$$H_{mj} = \frac{2}{(1-x_{mj}^2)\,[P_m'(x_{mj})]^2}, \tag{4.6.13}$$

while (4.6.10) and (4.6.11) become[1])

$$|R(u)| \leqq \frac{2^{2m+1}(m!)^4}{(2m+1)\,[(2m)!]^3} M_{2m}, \tag{4.6.14}$$

$$R(u) = \frac{2^{2m+1}(m!)^4}{(2m+1)\,[(2m)!]^3} u^{(2m)}(\xi), \quad (-1<\xi<1). \tag{4.6.15}$$

[1]) Remember that $\Gamma\left(m+\frac{1}{2}\right) = \frac{(2m)!}{2^{2m}\,m!}\sqrt{\pi}$ [see (3.2.7)].

Another interesting subcase is that in which

$$\lambda = 1 \qquad \left(\text{that is } \alpha = \beta = \frac{1}{2}\right).$$

It is known that the ultraspherical polynomials $P_m^{(1)}(x)$ coincide with *the Tchebychef polynomials of second kind* [see (3.5.7) and (3.5.10)]:

$$V_m(x) = \frac{\sin[(m+1)\arccos x]}{\sqrt{1-x^2}}, \tag{4.6.16}$$

whose zeros $x_{mi}^{(1)}$ are given by

$$x_{mi}^{(1)} = \cos\frac{i\pi}{m+1}, \qquad (i = 1, 2, \ldots, m).\,^{1)} \tag{4.6.17}$$

Therefore (4.6.1) becomes

$$\boxed{\int_{-1}^{1} \sqrt{1-x^2}\, u(x)\, dx = \sum_{i=1}^{m} H_{mi}^{(1)}\, u\left(\cos\frac{i\pi}{m+1}\right) + R(u)}, \tag{4.6.18}$$

with the coefficients expressed by

$$H_{mj}^{(1)} = \frac{\pi}{m+1}\sin^2\frac{j\pi}{m+1} \tag{4.6.19}$$

as can be found starting from (4.6.6) and carrying out an easy calculation.

The formulae (4.6.10) and (4.6.11) then reduce to

$$|R(u)| \leq \frac{\pi}{2^{2m+1}(2m)!} M_{2m}, \tag{4.6.20}$$

$$R(u) = \frac{\pi}{2^{2m+1}(2m)!} u^{(2m)}(\xi), \qquad (-1 < \xi < 1). \tag{4.6.21}$$

The second particular case of the Gauss-Jacobi formulae which we wish to examine is that one in which

$$\alpha = \beta = -\frac{1}{2}.$$

Remember that we have [see (3.5.4) and (3.5.11)]:

$$P_m^{\left(-\frac{1}{2}, -\frac{1}{2}\right)}(x) = \frac{(2m-1)!!}{(2m)!!} T_m(x) \tag{4.6.22}$$

where $T_m(x)$ are the *Tchebychef polynomials of first kind* defined by

$$T_m(x) = \cos(m \arccos x) \tag{4.6.23}$$

whose zeros are

$$x_{mi}^{\left(-\frac{1}{2}, -\frac{1}{2}\right)} = \cos\frac{2i-1}{2m}\pi, \qquad (i = 1, 2, \ldots, m).\,^{2)} \tag{4.6.24}$$

1) Here the nodes $x_{mi}^{(1)}$ are written in decreasing order.

2) In this case too the nodes are written in decreasing order.

The coefficients of the corresponding quadrature formula can be obtained from (4.5.6), using (4.6.22), (4.6.23) and (4.6.24). We find

$$H_{mj}^{\left(-\frac{1}{2},-\frac{1}{2}\right)} = \frac{\left[\Gamma\left(m+\frac{1}{2}\right)\right]^2}{m!\,\Gamma(m)\sin^2\frac{2j-1}{2m}\pi}\left[\frac{(2m-1)!!}{(2m)!!}T'_m\left(\cos\frac{2j-1}{2m}\pi\right)\right]^{-2}$$

and carrying out all the calculations

$$H_{mj}^{\left(-\frac{1}{2},-\frac{1}{2}\right)} = \frac{\pi}{m},$$

so that (4.5.2) becomes the classical *Tchebychef formula*

$$\boxed{\int_{-1}^{1}\frac{u(x)}{\sqrt{1-x^2}}\,dx = \frac{\pi}{m}\sum_{i=1}^{m}u\left(\cos\frac{2i-1}{2m}\pi\right)+R(u)}\ .^{1)} \qquad (4.6.25)$$

The reader will easily see that, in this case too, $\Phi(x)$ is an even function, whence (4.6.9) holds, while (4.5.18) and (4.5.19) become

$$|R(u)| \leqq \frac{\pi}{2^{2m-1}\,(2m)!}M_{2m}, \qquad (4.6.26)$$

$$R(u) = \frac{\pi}{2^{2m-1}\,(2m)!}u^{(2m)}(\xi), \quad (-1<\xi<1). \qquad (4.6.27)$$

Another important particular case is that in which we have

$$\alpha = \frac{1}{2}, \quad \beta = -\frac{1}{2}.$$ [2]

Remembering that [see (3.5.12)]:

$$P_m^{\left(\frac{1}{2},-\frac{1}{2}\right)}(x) = \frac{(2m-1)!!}{(2m)!!}\,\frac{\sin\left[\left(m+\frac{1}{2}\right)\arccos x\right]}{\sin\left(\frac{1}{2}\arccos x\right)},$$

we see that the nodes $x_{mi}^{\left(\frac{1}{2},-\frac{1}{2}\right)}$ are given (in decreasing order) by

$$x_{mi}^{\left(\frac{1}{2},-\frac{1}{2}\right)} = \cos\frac{2i\pi}{2m+1}, \quad (i=1,2,\ldots,m).$$

[1]) This formula can also be obtained in another way (see § 4.12).

[2]) The opposite case $\alpha = -\frac{1}{2}$, $\beta = \frac{1}{2}$ reduces to this, operating on the integral to be calculated, the substitution $x = -\xi$.

The quadrature formula is written

$$\boxed{\int_{-1}^{1} \sqrt{\frac{1-x}{1+x}}\, u(x)\, dx = \sum_{i=1}^{m} H_{mi}^{\left(\frac{1}{2}, -\frac{1}{2}\right)} u\left(\cos \frac{2\, i\, \pi}{2\, m+1}\right) + R(u)}\,, \qquad (4.6.28)$$

with the coefficients easily deduced from (4.5.6); carrying out the calculations we find

$$H_{mi}^{\left(\frac{1}{2}, -\frac{1}{2}\right)} = \frac{4\, \pi}{2\, m+1} \sin^2 \frac{i\, \pi}{2\, m+1}\,. \qquad (4.6.29)$$

From (4.5.15) we then obtain

$$\int_{-1}^{1} \Phi(x)\, dx = \frac{\pi}{2^{2m}\,(2\, m)!} \qquad (4.6.30)$$

from which the particular cases of (4.5.18) and (4.5.19) follow; the calculation is left to the reader.

4.7. The Bouzitat formulae of the first kind. The Bouzitat elementary quadrature formulae of the first kind are relative to the interval $[-1, 1]$ and to the weight $g(x) = (1-x)^{\alpha}\,(1+x)^{\beta}$ (with $\alpha > -1$, $\beta > -1$) and have the following form, *under the hypothesis* $u(x) \in A\; C^{2m}[-1, 1]$:

$$\int_{-1}^{1} (1-x)^{\alpha}\,(1+x)^{\beta}\, u(x)\, dx = (\beta+1)\, B_0\, u(-1) + \sum_{i=1}^{m} B_i\, u(x_i) + R(u)$$

with $m+1$ nodes $-1, x_1, x_2, \ldots, x_m$ and under the condition that $R(u) = 0$ *when* $u(x)$ *is a polynomial of degree* $\leq 2\,m$. From § 2.5 it can be seen that the procedure to obtain the preceding formula is equivalent to a Gauss problem with $n = 2\,m+1$, $E = \frac{d^{2m+1}}{dx^{2m+1}}$, $p = n-1 = 2\,m$. Therefore we must consider the boundary problem

$$\frac{d^{2m+1}u}{dx^{2m+1}} = 0\,, \quad u(-1) = u(x_1) = u(x_2) = \cdots = u(x_m) = 0\,,$$

which has m linearly independent solutions

$$U_r(x) = x^r\,(1+x) \prod_{i=1}^{m} (x - x_i)\,, \qquad (r = 0, 1, \ldots, m-1)\,.$$

On the basis of theor. 2.5.I, the formula exists if and only if the nodes x_i are chosen in such a way as to satisfy

$$\int_{-1}^{1} (1-x)^{\alpha}\,(1+x)^{\beta+1}\, x^r \prod_{i=1}^{m} (x - x_i)\, dx = 0\,, \qquad (r = 0, 1, \ldots, m-1)\,,$$

that is (see § 3.4) if and only if the nodes $x_1, x_2, \ldots, x_m$ *are the zeros* $x_{m1}^{(\alpha, \beta+1)}$, $x_{m2}^{(\alpha, \beta+1)}, \ldots, x_{mm}^{(\alpha, \beta+1)}$ *of the Jacobi polynomials* $P_m^{(\alpha, \beta+1)}(x)$.

With these nodes the formula is unique since $(m+1)(n-p)-n+q = (m+1)\cdot 1-(2m+1)+m=0$ (see theor. 2.5.I).

We therefore will write again the quadrature formula in the following way

$$\boxed{\begin{aligned}&\int_{-1}^{1}(1-x)^\alpha(1+x)^\beta u(x)\,dx\\ &\qquad=(\beta+1)\,B_{m0}^{(\alpha,\beta)}\,u(-1)+\sum_{i=1}^{m}B_{mi}^{(\alpha,\beta)}\,u(x_{mi}^{(\alpha,\beta+1)})+R(u)\end{aligned}}\,. \tag{4.7.1}$$

Now let us compute the coefficients $B_{m0}^{(\alpha,\beta)}$, $B_{mi}^{(\alpha,\beta)}$ and the remainder $R(u)$ by a procedure analogous to that carried out for the Gauss-Jacobi formulae (see § 4.5).

If in (4.7.1) we put

$$u(x)=\left[\frac{d}{dx}P_m^{(\alpha,\beta+1)}(x)\right]^{-1}_{x=x_{mj}^{(\alpha,\beta+1)}}\frac{(1+x)\,P_m^{(\alpha,\beta+1)}(x)}{(1+x_{mj}^{(\alpha,\beta+1)})\,(x-x_{mj}^{(\alpha,\beta+1)})}$$

and we consider that $R(u)=0$, $u(-1)=0$, $u(x_{mi}^{(\alpha,\beta+1)})=\delta_{ij}$, we obtain immediately

$$B_{mj}^{(\alpha,\beta)}=\left[\frac{d}{dx}P_m^{(\alpha,\beta+1)}(x)\right]^{-1}_{x=x_{mj}^{(\alpha,\beta+1)}}\times$$

$$\times\frac{1}{1+x_{mj}^{(\alpha,\beta+1)}}\int_{-1}^{1}(1-x)^\alpha(1+x)^{\beta+1}\frac{P_m^{(\alpha,\beta+1)}(x)}{x-x_{mj}^{(\alpha,\beta+1)}}\,dx\,,$$

$$(j=1,2,\ldots,m)\,,$$

or using the expressions (4.5.3) and (4.5.6) of the Christoffel numbers:

$$B_{mj}^{(\alpha,\beta)}=\frac{H_{mj}^{(\alpha,\beta+1)}}{1+x_{mj}^{(\alpha,\beta+1)}}$$

$$=\frac{2^{\alpha+\beta+2}\,\Gamma(\alpha+m+1)\,\Gamma(\beta+m+2)}{m!\,\Gamma(\alpha+\beta+m+2)\,(1+x_{mj}^{(\alpha,\beta+1)})\,[1-(x_{mj}^{(\alpha,\beta+1)})^2]}\left[\frac{d}{dx}P_m^{(\alpha,\beta+1)}(x)\right]^{-2}_{x=x_{mj}^{(\alpha,\beta+1)}}. \tag{4.7.2}$$

This formula can be transformed so as to let the polynomial $P_m^{(\alpha,\beta)}(x)$ occur in it explicitly; it suffices to consider that

$$\left[\frac{d}{dx}P_m^{(\alpha,\beta+1)}(x)\right]_{x=x_{mj}^{(\alpha,\beta+1)}}=\frac{2(\beta+m+1)}{1-(x_{mj}^{(\alpha,\beta+1)})^2}P_m^{(\alpha,\beta)}(x_{mj}^{(\alpha,\beta+1)})\ ^{1)}$$

to obtain

$$B_{mj}^{(\alpha,\beta)}=\frac{2^{\alpha+\beta}\,\Gamma(\alpha+m+1)\,\Gamma(\beta+m+1)\,(1-x_{mj}^{(\alpha,\beta+1)})}{m!\,(\beta+m+1)\,\Gamma(\alpha+\beta+m+2)\,[P_m^{(\alpha,\beta)}(x_{mj}^{(\alpha,\beta+1)})]^2}\,,$$

$$(j=1,2,\ldots,m)\,. \tag{4.7.3}$$

[1]) This formula follows from (3.4.23), (3.4.29) substituting β by $\beta+1$, putting $x=x_{mj}^{(\alpha,\beta+1)}$ and then eliminating $P_{m-1}^{(\alpha,\beta+1)}(x_{mj}^{(\alpha,\beta+1)})$.

Analogously, if we put

$$u(x) = \frac{P_m^{(\alpha,\beta+1)}(x)}{P_m^{(\alpha,\beta+1)}(-1)}$$

in (4.7.1) we obtain

$$(\beta + 1)\, B_{m0}^{(\alpha,\beta)} = \frac{1}{P_m^{(\alpha,\beta+1)}(-1)} \int_{-1}^{1} (1-x)^\alpha (1+x)^\beta P_m^{(\alpha,\beta+1)}(x)\, dx\,. \tag{4.7.4}$$

We may transform the right member, using the well known formulae [see (3.4.5), (3.4.34)]:

$$P_m^{(\alpha,\beta)}(-1) = (-1)^m \binom{\beta + m}{m},$$

$$\sum_{k=0}^{m} (-1)^k \frac{(\alpha+\beta+2k+1)\,\Gamma(\alpha+\beta+k+1)}{\Gamma(\alpha+k+1)} P_k^{(\alpha,\beta)}(x) = (-1)^m \frac{\Gamma(\alpha+\beta+m+2)}{\Gamma(\alpha+m+1)} P_m^{(\alpha,\beta+1)}(x)\,.$$

From the latter we deduce

$$\frac{(\alpha+\beta+1)\,\Gamma(\alpha+\beta+1)}{\Gamma(\alpha+1)} h_0(\alpha,\beta) = \frac{(-1)^m\,\Gamma(\alpha+\beta+m+2)}{\Gamma(\alpha+m+1)} \int_{-1}^{1} (1-x)^\alpha (1+x)^\beta P^{(\alpha,\beta+1)}(x)\, dx$$

by multiplying both members by $(1-x)^\alpha (1+x)^\beta$ and integrating from -1 to 1, and furthermore:

$$\int_{-1}^{1} (1-x)^\alpha (1+x)^\beta P_m^{(\alpha,\beta+1)}(x)\, dx = \frac{(-1)^m 2^{\alpha+\beta+1}\,\Gamma(\beta+1)\,\Gamma(\alpha+m+1)}{\Gamma(\alpha+\beta+m+2)}\,.$$

Then we easily derive

$$B_{m0}^{(\alpha,\beta)} = \frac{2^{\alpha+\beta+1}\, m!\,[\Gamma(\beta+1)]^2\,\Gamma(\alpha+m+1)}{\Gamma(\beta+m+2)\,\Gamma(\alpha+\beta+m+2)} \tag{4.7.5}$$

from (4.7.4). [1])

In order to apply (4.1.3) to obtain the expression of the remainder $R(u)$, we must get the solutions $\varphi_1(x), \varphi_2(x), \ldots, \varphi_{m+1}(x)$ of the differential equation $E^*(\varphi) = g$ which is now $\varphi^{(2m+1)}(x) = -(1-x)^\alpha (1+x)^\beta$. Having already computed the coefficients $B_{m0}^{(\alpha,\beta)}$, $B_{mi}^{(\alpha,\beta)}$ of the quadrature formula, we may

[1]) Note that the right member of (4.7.5) can be obtained from the right member of (4.7.3), writing -1 instead of $x_{mj}^{(\alpha,\beta+1)}$.

apply theor. 2.4.I which yields

$$\varphi_i(x) = -\int_{-1}^{x} \frac{(x-\xi)^{2m}}{(2m)!}(1-\xi)^\alpha (1+\xi)^\beta \, d\xi + (\beta+1)\, B_{m0}^{(\alpha,\beta)} \frac{(x+1)^{2m}}{(2m)!}$$

$$+ \sum_{j=1}^{i} B_{mj}^{(\alpha,\beta)} \frac{(x - x_{mj}^{(\alpha,\beta+1)})^{2m}}{(2m)!}, \quad (i = 1, 2, \ldots, m+1), \text{[1]} \quad (4.7.6)$$

since $K(x, \xi) = \dfrac{(x-\xi)^{2m}}{(2m)!}$.

We may then assert

$$R(u) = \int_{-1}^{1} \Phi(x)\, u^{(2m+1)}(x)\, dx$$

$$= \sum_{i=0}^{m} \int_{x_{mi}^{(\alpha,\beta+1)}}^{x_{m,i+1}^{(\alpha,\beta+1)}} \varphi_{i+1}(x)\, u^{(2m+1)}(x)\, dx, \quad (x_{m0}^{(\alpha,\beta+1)} = -1,\ x_{m,m+1}^{(\alpha,\beta+1)} = 1), \quad (4.7.7)$$

with $\varphi_i(x)$ expressed by (4.7.6), and

$$\Phi(x) = \varphi_{i+1}(x) \quad \text{for} \quad x_{mi}^{(\alpha,\beta+1)} < x < x_{m,i+1}^{(\alpha,\beta+1)}, \quad (i = 0, 1, \ldots, m). \quad (4.7.8)$$

We shall now give two theorems similar to those of § 4.5.

Theorem 4.7.I. *The influence-function $\Phi(x)$ defined by (4.7.8) [together with (4.7.6)] belongs to the class $C^{2m-1}[-1, 1]$ and is positive inside the interval.*

Proof. The proof is similar to that one of theor. 4.5.I. Writing down the formulae that we obtain by differentiating k times (with $0 \leqq k \leqq 2m$) the formulae (4.7.6), (4.7.8), we soon deduce that

$$\left.\begin{array}{l} \Phi^{(k)}(-1) = 0, \quad (k = 0, 1, \ldots, 2m-1); \quad \Phi^{(k)}(1) = 0, \\ \qquad\qquad\qquad\qquad\qquad\qquad (k = 0, 1, \ldots, 2m), \\ (-1)^k \Phi^{(k)}(x) > 0 \quad \text{for} \quad x_{mm}^{(\alpha,\beta+1)} < x < 1, \\ \qquad\qquad\qquad\qquad\qquad\qquad (k = 0, 1, \ldots, 2m), \\ \Phi(x), \Phi'(x), \ldots, \Phi^{(2m-1)}(x) \quad \text{continuous in } [-1, 1], \end{array}\right\} \quad (4.7.9)$$

while $\Phi^{(2m)}(x)$ has discontinuities of first kind at the points $x_{m1}^{(\alpha,\beta+1)}, \ldots, x_{mm}^{(\alpha,\beta+1)}$.

After that, using the fact that $\Phi^{(2m)}(x)$ is *decreasing* in each interval $(x_{mi}^{(\alpha,\beta+1)}, x_{m,i+1}^{(\alpha,\beta+1)})$, $(0 \leqq i \leqq m-1)$, we may deduce that $\Phi^{(2m-1)}(x)$ (which is zero for $x = -1$) can vanish at most once inside $[-1, x_{m1}^{(\alpha,\beta+1)}]$ and at most twice in each one of the $m-1$ intervals $[x_{mi}^{(\alpha,\beta+1)}, x_{m,i+1}^{(\alpha,\beta+1)}]$, $(i = 1, 2, \ldots, m-1)$. Therefore $\Phi^{(2m-1)}(x)$ vanishes at most $1 + 2(m-1) = 2m - 1$ times inside $[-1, 1]$ and there follows that $\Phi(x)$ (which is zero for $x = -1$ and $x = 1$) cannot vasnish inside $[-1, 1]$; we then see that this is here positive, using (4.7.9).

[1]) For $i = m+1$ we may also write $\varphi_{m+1}(x) = \displaystyle\int_{x}^{1} \frac{(x-\xi)^{2m}}{(2m)!}(1-\xi)^\alpha (1+\xi)^\beta \, d\xi$.

Theorem 4.7.II. *We have*

$$\int_{-1}^{1} \Phi(x)\, dx = \frac{2^{\alpha+\beta+2m+2}\, m!\, \Gamma(\alpha+m+1)\, \Gamma(\beta+m+2)\, \Gamma(\alpha+\beta+m+2)}{(2m+1)!\, (\alpha+\beta+2m+2)\, [\Gamma(\alpha+\beta+2m+2)]^2}\,. \tag{4.7.10}$$

Proof. Let us examine again (4.7.1) and substitute in it

$$u(x) = (1+x)\, [P_m^{(\alpha,\beta+1)}(x)]^2 = (1+x)\, [a_m x^m + \cdots]^2\,.$$

Using (4.7.7) we obtain

$$\int_{-1}^{1} (1-x)^\alpha (1+x)^{\beta+1} [P_m^{(\alpha,\beta+1)}(x)]^2\, dx = (2m+1)!\, a_m^2 \int_{-1}^{1} \Phi(x)\, dx\,,$$

from which, with a reasoning similar to that of the proof of theor. 4.5.II., we deduce (4.7.10), q.e.d.

Now we examine again the expression (4.7.7) for the remainder $R(u)$. *Under only the hypothesis* $u(x) \in AC^{2m}[-1, 1]$ we establish a bound analogous to that of (4.5.16) and (4.5.17), namely:

$$|R(u)| \leqq \Phi(x_0)\, V_{2m} \tag{4.7.11}$$

where x_0 is the only point inside $[-1, 1]$ where $\Phi'(x)$ vanishes and V_{2m} is the total variation in $[-1, 1]$ of the absolutely continuous function $u^{(2m)}(x)$.

Assuming the hypothesis *that* $u^{(2m+1)}(x)$ *be bounded in* $[-1, 1]$, and putting $M_{2m+1} = \sup\limits_{-1 \leqq x \leqq 1} |u^{(2m+1)}(x)|$, (4.7.7) yields by Theor. 4.7.I and 4.7.II:

$$|R(u)| \leqq \frac{2^{\alpha+\beta+2m+2}\, m!\, \Gamma(\alpha+m+1)\, \Gamma(\beta+m+2)\, \Gamma(\alpha+\beta+m+2)}{(2m+1)!\, (\alpha+\beta+2m+2)\, [\Gamma(\alpha+\beta+2m+2)]^2}\, M_{2m+1}\,. \tag{4.7.12}$$

Finally, *if* $u^{(2m+1)}(x)$ *is continuous in* $[-1, 1]$, we obtain

$$R(u) = \frac{2^{\alpha+\beta+2m+2}\, m!\, \Gamma(\alpha+m+1)\, \Gamma(\beta+m+2)\, \Gamma(\alpha+\beta+m+2)}{(2m+1)!\, (\alpha+\beta+2m+2)\, [\Gamma(\alpha+\beta+2m+2)]^2}\, u^{(2m+1)}(\xi)\,,$$

$$(-1 < \xi < 1)\,, \tag{4.7.13}$$

applying the mean value theorem and using theor. 4.7.I.

The particular case of (4.7.1) in which $\alpha = \beta = 0$ is important for we obtain the classical *Radau formula*

$$\boxed{\int_{-1}^{1} u(x)\, dx = B_{m0}^{(0,0)}\, u(-1) + \sum_{i=1}^{m} B_{mi}^{(0,0)}\, u(x_{mi}^{(0,1)}) + R(u)}\,, \tag{4.7.14}$$

which is exact when $u(x)$ is a polynomial of degree $\leqq 2m$ and in which the nodes $x_{mi}^{(0,1)}$ are zeros of the Jacobi polynomial $P_m^{(0,1)}(x)$. Equations (4.7.2),

(4.7.3), (4.7.5), (4.7.12) and (4.7.13) become

$$B_{mj}^{(0,0)} = \frac{4}{(1 + x_{mj}^{(0,1)})\,[1 - (x_{mj}^{(0,1)})^2]} \left[\frac{d}{dx} P_m^{(0,1)}(x)\right]_{x = x_{mj}^{(0,1)}}^{-2}$$

$$= \frac{1 - x_{mj}^{(0,1)}}{(m+1)^2\,[P_m(x_{mj}^{(0,1)})]^2}\,,^{1)} \qquad (j = 1, 2, \ldots, m)$$

$$B_{m0}^{(0,0)} = \frac{2}{(m+1)^2}\,,$$

$$\left|R(u)\right| \leqq \frac{2^{2m+1}\,(m+1)\,(m!)^4}{[(2m+1)!]^3}\, M_{2m+1}\,,$$

$$R(u) = \frac{2^{2m+1}\,(m+1)\,(m!)^4}{[(2m+1)!]^3}\, u^{(2m+1)}(\xi)\,, \qquad (-1 < \xi < 1)\,.$$

4.8. The Bouzitat formulae of the second kind. These formulae are relative to the interval $[-1, 1]$ and to the weight $g(x) = (1-x)^\alpha\,(1+x)^\beta$ (with $\alpha > -1$, $\beta > -1$), under the hypothesis $u(x) \in AC^{2m+1}[-1, 1]$. They have the form

$$\int_{-1}^{1} (1-x)^\alpha\,(1+x)^\beta\,u(x)\,dx = (\beta+1)\,C_0\,u(-1) + \sum_{i=1}^{m} C_i\,u(x_i)$$
$$+ (\alpha+1)\,C_{m+1}\,u(1) + R(u)\,,$$

with $m+2$ nodes $-1, x_1, x_2, \ldots, x_m, 1$ and *under the condition that* $R(u) = 0$ *when* $u(x)$ *is a polynomial of degree* $\leqq 2m+1$. Reasoning as in the preceding § 4.7, we see that we must consider the boundary problem

$$\frac{d^{2m+2}}{dx^{2m+2}} = 0\,, \qquad u(-1) = u(x_1) = \cdots = u(x_m) = u(1) = 0$$

which has m linearly independent solutions

$$U_r(x) = x^r\,(1 - x^2) \prod_{i=1}^{m} (x - x_i)\,, \qquad (r = 0, 1, \ldots, m-1)\,.$$

The formula can exist if and only if the equations

$$\int_{-1}^{1} (1-x)^{\alpha+1}\,(1+x)^{\beta+1}\,x^r \prod_{i=1}^{m} (x - x_i)\,dx = 0\,, \qquad (r = 0, 1, \ldots, m-1)\,,$$

are satisfied, that is if and only if the *nodes* $x_1, x_2, \ldots, x_m$ *are the zeros* $x_{m1}^{(\alpha+1,\beta+1)}$, $x_{m2}^{(\alpha+1,\beta+1)}, \ldots, x_{mm}^{(\alpha+1,\beta+1)}$ *of the Jacobi polynomials* $P_m^{(\alpha+1,\beta+1)}(x)$.

Therefore it is convenient to write again the formula in the following way

$$\boxed{\begin{aligned} &\int_{-1}^{1} (1-x)^\alpha\,(1+x)^\beta\,u(x)\,dx = (\beta+1)\,C_{m0}^{(\alpha,\beta)}\,u(-1) \\ &\quad + \sum_{i=1}^{m} C_{mi}^{(\alpha,\beta)}\,u(x_{mi}^{(\alpha+1,\beta+1)}) + (\alpha+1)\,C_{m,m+1}^{(\alpha,\beta)}\,u(1) + R(u) \end{aligned}}\,, \qquad (4.8.1)$$

1) Here $P_m(x) = P_m^{(0,0)}(x)$ denotes the Legendre polynomial of degree m.

noting that it is unique since (theor. 2.5.I) we have $(m + 2)(n - p) - n + q = (m + 2) \cdot 1 - (2m + 2) + m = 0$.

In order to compute the coefficients $C_{mi}^{(\alpha,\beta)}$ $(i = 1, 2, \ldots, m)$, it suffices to substitute in (4.8.1)

$$u(x) = \left[\frac{d}{dx} P_m^{(\alpha+1,\beta+1)}(x)\right]^{-1}_{x=x_{mj}^{(\alpha+1,\beta+1)}} \frac{(1-x^2) P_m^{(\alpha+1,\beta+1)}(x)}{[1-(x_{mj}^{(\alpha+1,\beta+1)})^2](x - x_{mj}^{(\alpha+1,\beta+1)})}$$

and to remember (4.5.3) and (4.5.6) to obtain

$$C_{mj}^{(\alpha,\beta)} = \frac{H_{mj}^{(\alpha+1,\beta+1)}}{1-(x_{mj}^{(\alpha+1,\beta+1)})^2} = \frac{2^{\alpha+\beta+3}\,\Gamma(\alpha+m+2)\,\Gamma(\beta+m+2)}{m!\,\Gamma(\alpha+\beta+m+3)\,[1-(x_{mj}^{(\alpha+1,\beta+1)})^2]^2} \times$$
$$\times \left[\frac{d}{dx} P_m^{(\alpha+1,\beta+1)}(x)\right]^{-2}_{x=x_{mj}^{(\alpha+1,\beta+1)}}, \quad (j = 1, 2, \ldots, m). \tag{4.8.2}$$

By the use of

$$[1-(x_{mj}^{(\alpha+1,\beta+1)})^2]\left[\frac{d}{dx} P_m^{(\alpha+1,\beta+1)}(x)\right]_{x=x_{mj}^{(\alpha+1,\beta+1)}} = -2(m+1) P_{m+1}^{(\alpha,\beta)}(x_{mj}^{(\alpha+1,\beta+1)}) \text{ [1]} \tag{4.8.3}$$

the formula (4.8.2) can be transformed into

$$C_{mj}^{(\alpha,\beta)} = \frac{2^{\alpha+\beta+1}\,\Gamma(\alpha+m+2)\,\Gamma(\beta+m+2)}{(m+1)(m+1)!\,\Gamma(\alpha+\beta+m+3)\,[P_{m+1}^{(\alpha,\beta)}(x_{mj}^{(\alpha+1,\beta+1)})]^2},$$
$$(j = 1, 2, \ldots, m). \tag{4.8.4}$$

In order to compute the coefficients $C_{m0}^{(\alpha,\beta)}$, $C_{m,m+1}^{(\alpha,\beta)}$ we substitute in (4.8.1):

$$u(x) = \frac{1-x}{2}\,\frac{P_m^{(\alpha+1,\beta+1)}(x)}{P_m^{(\alpha+1,\beta+1)}(-1)}, \qquad u(x) = \frac{1+x}{2}\,\frac{P_m^{(\alpha+1,\beta+1)}(x)}{P_m^{(\alpha+1,\beta+1)}(1)};$$

we find

$$(\beta+1)\,C_{m0}^{(\alpha,\beta)} = \frac{1}{2P_m^{(\alpha+1,\beta+1)}(-1)} \int_{-1}^{1} (1-x)^{\alpha+1}(1+x)^{\beta} P_m^{(\alpha+1,\beta+1)}(x)\,dx,$$

$$(\alpha+1)\,C_{m,m+1}^{(\alpha,\beta)} = \frac{1}{2P_m^{(\alpha+1,\beta+1)}(1)} \int_{-1}^{1} (1-x)^{\alpha}(1+x)^{\beta+1} P_m^{(\alpha+1,\beta+1)}(x)\,dx.$$

[1] If we write the linear differential equation of second order satisfied by $P_{m+1}^{(\alpha,\beta)}(x)$ [see (3.4.17)] and transform the two terms containing derivatives by means of the formula [see (3.4.21)] $\frac{d}{dx} P_{m+1}^{(\alpha,\beta)}(x) = \frac{1}{2}(\alpha+\beta+m+2) P_m^{(\alpha+1,\beta+1)}(x)$, we obtain $(1-x^2)\frac{d}{dx} P_m^{(\alpha+1,\beta+1)}(x) + [\beta - \alpha - (\alpha+\beta+2)x] P_m^{(\alpha+1,\beta+1)}(x) + 2(m+1) P_{m+1}^{(\alpha,\beta)}(x) = 0$, which with $x = x_{mj}^{(\alpha+1,\beta+1)}$ yields (4.8.3).

Transforming this formula by means of a procedure analogous to the one used in the preceding § 4.7, we easily obtain

$$C_{m0}^{(\alpha,\beta)} = \frac{2^{\alpha+\beta+1}\, m!\, [\Gamma(\beta+1)]^2\, \Gamma(\alpha+m+2)}{\Gamma(\beta+m+2)\, \Gamma(\alpha+\beta+m+3)}, \tag{4.8.5}$$

$$C_{m,m+1}^{(\alpha,\beta)} = \frac{2^{\alpha+\beta+1}\, m!\, [\Gamma(\alpha+1)]^2\, \Gamma(\beta+m+2)}{\Gamma(\alpha+m+2)\, \Gamma(\alpha+\beta+m+3)}.\ ^{1)} \tag{4.8.6}$$

In order to obtain, by means of (4.1.3), the expression for the remainder $R(u)$ we must get the solutions $\varphi_1(x), \varphi_2(x), \ldots, \varphi_{m+1}(x)$ of the differential equation $\varphi^{(2m+2)}(x) = (1-x)^\alpha (1+x)^\beta$. Having already determined the coefficients $C_{m0}^{(\alpha,\beta)}, C_{mi}^{(\alpha,\beta)}, C_{m,m+1}^{(\alpha,\beta)}$, we may apply theor. 2.4.I which yields

$$\varphi_i(x) = \int_{-1}^{x} \frac{(x-\xi)^{2m+1}}{(2m+1)!}(1-\xi)^\alpha (1+\xi)^\beta\, d\xi - (\beta+1)\, C_{m0}^{(\alpha,\beta)} \frac{(x+1)^{2m+1}}{(2m+1)!}$$

$$- \sum_{j=1}^{i-1} C_{mj}^{(\alpha,\beta)} \frac{(x - x_{mj}^{(\alpha+1,\beta+1)})^{2m+1}}{(2m+1)!},\ ^{2)} \quad (i = 1, 2, \ldots, m+1), \tag{4.8.7}$$

since $K(x,\xi) = \dfrac{(x-\xi)^{2m+1}}{(2m+1)!}$. Next we have

$$R(u) = \int_{-1}^{1} \Phi(x)\, u^{(2m+2)}(x)\, dx = \sum_{i=0}^{m} \int_{x_{mi}^{(\alpha+1,\beta+1)}}^{x_{m,i+1}^{(\alpha+1,\beta+1)}} \varphi_{i+1}(x)\, u^{(2m+2)}(x)\, dx,$$

$$(x_{m0}^{(\alpha+1,\beta+1)} = -1,\ x_{m,m+1}^{(\alpha+1,\beta+1)} = 1), \tag{4.8.8}$$

with

$$\Phi(x) = \varphi_{i+1}(x) \quad \text{for} \quad x_{mi}^{(\alpha+1,\beta+1)} < x < x_{m,i+1}^{(\alpha+1,\beta+1)}, \quad (i = 0, 1, \ldots, m). \tag{4.8.9}$$

Then we have two theorems analogous to those in § 4.5 and § 4.7.

Theorem 4.8.I. *The influence-function $\Phi(x)$ defined by (4.8.9) and (4.8.7) belongs to the class $C^{2m}[-1, 1]$ and is negative inside the interval.*

Proof. The proof is similar to that one of theor. 4.5.I (or 4.7.I). Writing down the formulae that we obtain by differentiating k times (with $0 \leqq k \leqq 2m+1$) the formulae (4.8.7), (4.8.9), we soon deduce that

$$\left.\begin{aligned} &\Phi^{(k)}(-1) = 0, \quad \Phi^{(k)}(1) = 0, \quad (k = 0, 1, \ldots, 2m), \\ &\Phi(x), \Phi'(x), \ldots, \Phi^{(2m)}(x) \quad \text{continuous in } [-1, 1], \end{aligned}\right\} \tag{4.8.10}$$

[1]) It is easy to verify that the right members of (4.8.5) and (4.8.6) can be obtained from the right member of (4.8.4) by writing -1 or 1 instead of $x_{mj}^{(\alpha+1,\beta+1)}$.

[2]) For $i = m+1$ we may also write

$$\varphi_{m+1}(x) = -\int_{x}^{1} \frac{(x-\xi)^{2m+1}}{(2m+1)!}(1-\xi)^\alpha (1+\xi)^\beta\, d\xi + (\alpha+1)\, C_{m,m+1}^{(\alpha,\beta)} \frac{(x-1)^{2m+1}}{(2m+1)!}.$$

while $\Phi^{(2m+1)}(x)$ has discontinuities of first kind at the points $x_{m1}^{(\alpha+1,\beta+1)}, \ldots, x_{mm}^{(\alpha+1,\beta+1)}$.

After that, using the fact that $\Phi^{(2m+1)}(x)$ is *increasing* in each interval $(x_{mi}^{(\alpha+1,\beta+1)}, x_{m,i+1}^{(\alpha+1,\beta+1)})$, $(0 \leqq i \leqq m)$, we may deduce that $\Phi^{(2m)}(x)$ (which is zero for $x = -1$ and $x = 1$) can vanish at most once inside $[-1, x_{m1}^{(\alpha+1,\beta+1)}]$ and $[x_{mm}^{(\alpha+1,\beta+1)}, 1]$ and at most twice in each one of the $m-1$ intervals $[x_{mi}^{(\alpha+1,\beta+1)}, x_{m,i+1}^{(\alpha+1,\beta+1)}]$, $(i = 1, 2, \ldots, m-1)$.

Therefore $\Phi^{(2m)}(x)$ vanishes at most $2 + 2\,(m-1) = 2\,m$ times inside $[-1, 1]$ and there follows that $\Phi(x)$ (which is zero for $x = -1$ and $x = 1$) cannot vanish inside $[-1, 1]$. Then the result $\Phi(x) < 0$ (for $-1 < x < 1$) derives from the fact (proved by the next theorem) that $\int_{-1}^{1} \Phi(x)\,dx < 0$.

Theorem 4.8.II. *We have*

$$\int_{-1}^{1} \Phi(x)\,dx = -\frac{2^{\alpha+\beta+2m+3}\, m!\, \Gamma(\alpha+m+2)\, \Gamma(\beta+m+2)\, \Gamma(\alpha+\beta+m+3)}{(2m+2)!\,(\alpha+\beta+2m+3)\,[\Gamma(\alpha+\beta+2m+3)]^2}\,. \tag{4.8.11}$$

Proof. Let us substitute

$$u(x) = (1-x^2)\,[P_m^{(\alpha+1,\beta+1)}(x)]^2 = (1-x^2)\,[a_m x^m + \cdots]^2$$

in (4.8.1). Bearing in mind (4.8.8) we obtain

$$\int_{-1}^{1} (1-x)^{\alpha+1}\,(1+x)^{\beta+1}\,[P_m^{(\alpha+1,\beta+1)}(x)]^2\,dx = -\,(2m+2)!\,a_m^2 \int_{-1}^{1} \Phi(x)\,dx\,,$$

from which, reasoning as in the proof of theor. 4.5.II, we obtain (4.8.11).

Now let us examine again the expression (4.8.8) for the remainder $R(u)$. *Under the sole hypothesis* $u(x) \in AC^{2m+1}[-1, 1]$ we immediately deduce a bound similar to that derived from (4.5.16) and (4.5.17):

$$|R(u)| \leqq -\,\Phi(x_0)\,V_{2m+1} \tag{4.8.12}$$

where x_0 is the only point inside $[-1, 1]$ where $\Phi'(x)$ vanishes and V_{2m+1} is the total variation in $[-1, 1]$ of the absolutely continuous function $u^{(2m+1)}(x)$. If we make the hypothesis *that* $u^{(2m+2)}(x)$ *be bounded in* $[-1, 1]$, and put $M_{2m+2} = \sup_{-1 \leqq x \leqq 1} |u^{(2m+2)}(x)|$, (4.8.8) yields:

$$|R(u)| \leqq \frac{2^{\alpha+\beta+2m+3}\, m!\, \Gamma(\alpha+m+2)\, \Gamma(\beta+m+2)\, \Gamma(\alpha+\beta+m+3)}{(2m+2)!\,(\alpha+\beta+2m+3)\,[\Gamma(\alpha+\beta+2m+3)]^2}\, M_{2m+2}\,, \tag{4.8.13}$$

by virtue of theorems 4.8.I and 4.8.II.

Finally *if* $u^{(2m+2)}(x)$ *is continuous in* $[-1, 1]$, we obtain:

$$R(u) = -\frac{2^{\alpha+\beta+2m+3}\, m!\, \Gamma(\alpha+m+2)\, \Gamma(\beta+m+2)\, \Gamma(\alpha+\beta+m+3)}{(2m+2)!\,(\alpha+\beta+2m+3)\,[\Gamma(\alpha+\beta+2m+3)]^2}\, u^{(2m+2)}(\xi)\,,$$

$$(-1 < \xi < 1)\,. \tag{4.8.14}$$

by applying the mean value theorem and theorem 4.8.I.

It is worthwhile to note the particular case $\alpha = \beta = \lambda - \frac{1}{2}$ of (4.8.1), for which the reader must see § 4.15, Problem 5. Here we confine ourselves to point out the important subcase $\alpha = \beta = 0$ $\left(\text{that is } \lambda = \frac{1}{2}\right)$ in which (4.8.1) becomes the classical *Lobatto formula*:

$$\boxed{\int_{-1}^{1} u(x)\,dx = C_{m0}^{(0,0)}\,u(-1) + \sum_{i=1}^{m} C_{mi}^{(0,0)}\,u(x_{mi}^{(1,1)}) + C_{m,m+1}^{(0,0)}\,u(1) + R(u)}\,, \tag{4.8.15}$$

which is exact when $u(x)$ is a polynomial of degree $\leqq 2m + 1$ and in which the nodes $x_{mi}^{(1,1)}$ are the zeros of the Jacobi polynomial $P_m^{(1,1)}(x)$, which are symmetric with respect to the point $x = 0$. Equations (4.8.2), (4.8.4), (4.8.5), (4.8.6), (4.8.13) and (4.8.14) become

$$C_{mj}^{(0,0)} = \frac{8\,(m+1)}{(m+2)\,[1-(x_{mj}^{(1,1)})^2]^2}\left[\frac{d}{dx}P_m^{(1,1)}(x)\right]_{x=x_{mj}^{(1,1)}}^{-2}$$

$$= \frac{2}{(m+1)\,(m+2)\,[P_{m+1}(x_{mj}^{(1,1)})]^2}\,,\ ^{1)} \qquad (j = 1, 2, \ldots, m)$$

$$C_{m0}^{(0,0)} = C_{m,m+1}^{(0,0)} = \frac{2}{(m+1)\,(m+2)}\,,$$

$$|R(u)| \leqq \frac{2^{2m+3}\,m!\,[(m+1)!]^2\,(m+2)!}{(2m+3)\,[(2m+2)!]^3}\,M_{2m+2}\,,$$

$$R(u) = -\frac{2^{2m+3}\,m!\,[(m+1)!]^2\,(m+2)!}{(2m+3)\,[(2m+2)!]^3}\,u^{(2m+2)}(\xi)\,, \qquad (-1 < \xi < 1)\,.$$

4.9. Gauss-Laguerre formulae. The Gauss-Laguerre elementary quadrature formulae are relative to the interval $[0, +\infty)$ and to the weight $g(x) = x^\alpha e^{-x}$ (with $\alpha > -1$) and have the following form

$$\boxed{\int_{0}^{+\infty} x^\alpha e^{-x}\,u(x)\,dx = \sum_{i=1}^{m} H_{mi}^{(\alpha)}\,u(x_{mi}^{(\alpha)}) + R(u)}\,, \tag{4.9.1}$$

with m nodes $x_{m1}^{(\alpha)}, \ldots, x_{mm}^{(\alpha)}$, under the condition that $R(u) = 0$ *when* $u(x)$ *is a polynomial of degree* $\leqq 2m - 1$ $\left(\text{that is that we have } E = \frac{d^{2m}}{dx^{2m}}\right)$.

We shall obtain (4.9.1) under hypotheses of the type (2.2.11), (2.2.12) keeping into account that, for the solutions $u_i(x)$, $v_i(x)$ of $E(u) = 0$ and

[1]) Here $P_{m+1}(x) = P_{m+1}^{(0,0)}(x)$ is the Legendre polynomial of degree $m + 1$. Note that there results $C_{mj}^{(0,0)} = C_{m,m-j+1}^{(0,0)}$.

$E^*(v) = 0$, we may put (see § 1.5, Problem 1):

$$u_i(x) = \frac{x^{i-1}}{(i-1)!}, \quad v_i(x) = (-1)^{2m-i}\frac{x^{2m-i}}{(2m-i)!}, \quad (i = 1, 2, \ldots, 2m). \tag{4.9.2}$$

Then we write (2.2.11) and (2.2.12) as follows

$$x^\alpha e^{-x}\frac{x^{i-1}}{(i-1)!} \in L\,[0, +\infty)\,,$$

$$(-1)^{2m-i}\frac{x^{2m-i}}{(2m-i)!}u^{(2m)}(x)\int_x^{+\infty}\xi^\alpha e^{-\xi}\frac{\xi^{i-1}}{(i-1)!}d\xi \in L\,[0, +\infty)\,,$$

$$(i = 1, 2, \ldots, 2m);$$

the first formulae are satisfied, while the second ones, considering that

$$\int_x^{+\infty}\xi^{\alpha+i-1}e^{-\xi}d\xi \sim x^{\alpha+i-1}e^{-x}, \quad (\text{for } x \to +\infty)\,,$$

reduce to the sole condition

$$x^{\alpha+2m-1}e^{-x}u^{(2m)}(x) \in L\,[0, +\infty)\,. \tag{4.9.3}$$

Now we consider how nodes must be chosen in order that the *Gauss problem* expressed by (4.9.1) have solution [see § 2.5].

The corresponding boundary problem is

$$\frac{d^{2m}}{dx^{2m}} = 0\,; \quad u(x_{mi}^{(\alpha)}) = 0\,, \quad (i = 1, 2, \ldots, m)\,;$$

it has the following m linearly independent solutions $U_r(x) = x^r \prod_{i=1}^{m}(x - x_{mi}^{(\alpha)})$ $(r = 0, 1, \ldots, m-1)$ and therefore (4.9.1) holds only if

$$\int_0^{+\infty}x^\alpha e^{-x}x^r\prod_{i=1}^{m}(x - x_{mi}^{(\alpha)})\,dx = 0\,, \quad (r = 0, 1, \ldots, m-1)\,,$$

that is (see § 3.6) only if the polynomial $\prod_{i=1}^{m}(x - x_{mi}^{(\alpha)})$ coincides, apart from a constant factor, with the Laguerre polynomial $L_m^{(\alpha)}(x)$. Then the *nodes* $x_{m1}^{(\alpha)}, \ldots, x_{mm}^{(\alpha)}$ *must coincide with the zeros of the Laguerre polynomial* $L_m^{(\alpha)}(x)$.

Choosing the nodes in this way, the formula (4.9.1) is unique since, on the basis of notations of § 2.5, we have

$$m\,(n-p) - n + q = m \cdot 1 - 2m + m = 0.$$

This is confirmed by the fact that if we put in (4.9.1)

$$u(x) = \left[\frac{d}{dx}L_m^{(\alpha)}(x)\right]_{x=x_{mj}^{(\alpha)}}^{-1}\frac{L_m^{(\alpha)}(x)}{x - x_{mj}^{(\alpha)}}\,,$$

we immediately find

$$H_{mj}^{(\alpha)} = \left[\frac{d}{dx}L_m^{(\alpha)}(x)\right]_{x=x_{mj}^{(\alpha)}}^{-1}\int_0^{+\infty}x^\alpha e^{-x}\frac{L_m^{(\alpha)}(x)}{x - x_{mj}^{(\alpha)}}dx\,, \quad (j = 1, 2, \ldots, m)\,. \tag{4.9.4}$$

To compute this integral we must observe that, putting $y = x_{mi}^{(\alpha)}$ in the Christoffel-Darboux summation formula [see (3.6.9)], we obtain

$$\sum_{k=0}^{m-1} \frac{L_k^{(\alpha)}(x)\, L_k^{(\alpha)}(x_{mj}^{(\alpha)})}{\binom{\alpha+k}{k}} = \frac{m+1}{\binom{\alpha+m}{m}} \frac{L_m^{(\alpha)}(x)\, L_{m+1}^{(\alpha)}(x_{mj}^{(\alpha)})}{x - x_{mj}^{(\alpha)}},$$

from which we deduce

$$\Gamma(\alpha+1) = \frac{m+1}{\binom{\alpha+m}{m}} L_{m+1}^{(\alpha)}(x_{mj}^{(\alpha)}) \int_0^{+\infty} x^\alpha e^{-x} \frac{L_m^{(\alpha)}(x)}{x - x_{mj}^{(\alpha)}}\, dx, \tag{4.9.5}$$

by multiplying both members by $x^\alpha e^{-x}$ and integrating from 0 to $+\infty$.

Moreover we have, using (3.6.12):

$$(m+1)\, L_{m+1}^{(\alpha)}(x_{mj}^{(\alpha)}) = x_{mj}^{(\alpha)} \left[\frac{d}{dx} L_m^{(\alpha)}(x)\right]_{x=x_{mj}^{(\alpha)}} \tag{4.9.6}$$

and from (4.9.4), (4.9.5) and (4.9.6) we then conclude that

$$H_{mj}^{(\alpha)} = \frac{\Gamma(\alpha+m+1)}{m!\, x_{mj}^{(\alpha)}} \left[\frac{d}{dx} L_m^{(\alpha)}(x)\right]_{x=x_{mj}^{(\alpha)}}^{-2}. \tag{4.9.7}$$

In order to apply (4.1.3) to obtain the expression of the remainder $R(u)$, we must get the solutions $\varphi_0(x), \varphi_1(x), \ldots, \varphi_m(x)$ of the differential equation $E^*(\varphi) = g$ which is now $\varphi^{(2m)}(x) = x^\alpha e^{-x}$. They are supplied by theorem 2.4.I from which we deduce

$$\varphi_i(x) = \int_0^x \frac{(x-\xi)^{2m-1}}{(2m-1)!} \xi^\alpha e^{-\xi}\, d\xi - \sum_{j=1}^{i} H_{mj}^{(\alpha)} \frac{(x - x_{mj}^{(\alpha)})^{2m-1}}{(2m-1)!}, \quad (i = 0, 1, \ldots, m)\,[1] \tag{4.9.8}$$

after which we have

$$R(u) = \int_0^{+\infty} \Phi(x)\, u^{(2m)}(x)\, dx = \sum_{i=0}^{m} \int_{x_{mi}^{(\alpha)}}^{x_{m,i+1}^{(\alpha)}} \varphi_i(x)\, u^{(2m)}(x)\, dx,$$

$$(x_{m0}^{(\alpha)} = 0,\ x_{m,m+1}^{(\alpha)} = +\infty), \tag{4.9.9}$$

where the influence-function $\Phi(x)$ is given by

$$\Phi(x) = \varphi_i(x) \quad \text{for} \quad x_{mi}^{(\alpha)} < x < x_{m,i+1}^{(\alpha)}, \quad (i = 0, 1, \ldots, m). \tag{4.9.10}$$

Now we prove two theorems analogous to those of the preceding § 4.5, 4.7 and 4.8.

[1]) For $i = m$ we may also write $\varphi_m(x) = -\int_x^{+\infty} \frac{(x-\xi)^{2m-1}}{(2m-1)!} \xi^\alpha e^{-\xi}\, d\xi$.

Theorem 4.9.I. *The influence-function $\Phi(x)$ defined by (4.9.10) [together with (4.9.8)] belongs to the class $C^{2m-2}[0, +\infty)$ and is positive inside such interval.*

Proof. The proof is similar to that one of theor. 4.5.I (or 4.7.I, 4.8.I). Writing down the formulae that we obtain differentiating k times (with $0 \leqq k \leqq 2m-1$) the formulae (4.9.8), (4.9.10), we soon deduce that

$$\Phi^{(k)}(0) = \Phi^{(k)}(+\infty) = 0$$

$$\Phi^{(k)}(x) > 0 \quad \text{for} \quad 0 < x < x_{m1}^{(\alpha)}, \ (-1)^k \Phi^{(k)}(x) > 0 \quad \text{for} \quad x_{mm}^{(\alpha)} < x < +\infty$$
$$(k = 0, 1, \ldots, 2m-1) \tag{4.9.11}$$

$$\Phi(x), \Phi'(x), \ldots, \Phi^{(2m-2)}(x) \text{ continuous in } [0, +\infty),$$

while $\Phi^{(2m-1)}(x)$ has discontinuities of first kind at the points $x_{m1}^{(\alpha)}, \ldots, x_{mm}^{(\alpha)}$.

After that, using the fact that $\Phi^{(2m-1)}(x)$ is *increasing* in each interval $(x_{mi}^{(\alpha)}, x_{m,i+1}^{(\alpha)})$, we can deduce that $\Phi^{(2m-2)}(x)$ vanishes at most twice in each one of the $m-1$ intervals $[x_{mi}^{(\alpha)}, x_{m,i+1}^{(\alpha)}]$, $(i = 1, 2, \ldots, m-1)$. Therefore $\Phi^{(2m-2)}(x)$ vanishes at most $2m-2$ times inside $[0, +\infty)$ and there follows that $\Phi(x)$ (which is zero for $x = 0$ and $x = +\infty$) cannot vanish inside $[-1, 1]$ and is positive, by virtue of (4.9.11).

Theorem 4.9.II. *We have*

$$\int_0^{+\infty} \Phi(x)\, dx = \frac{m!\, \Gamma(\alpha + m + 1)}{(2m)!}. \tag{4.9.12}$$

Proof. Let us examine again (4.9.1) and substitute in it

$$u(x) = [L_m^{(\alpha)}(x)]^2 = (a_m x^m + \cdots)^2 .$$

Using (4.9.9) we obtain

$$\int_0^{+\infty} x^\alpha e^{-x} [L_m^{(\alpha)}(x)]^2\, dx = (2m)!\, a_m^2 \int_0^{+\infty} \Phi(x)\, dx ,$$

and therefore (see § 3.6) remembering that the value of the integral in the first member is $\dfrac{\Gamma(\alpha + m + 1)}{m!}$ and that $a_m = \dfrac{(-1)^m}{m!}$ we conclude that effectively (4.9.12) holds, q.e.d.

From (4.9.9) and from theor. 4.9.I there follows that, *if $u^{(2m)}(x)$ has constant sign in $[0, +\infty)$, $R(u)$ has the same sign.*

Moreover, assuming the hypothesis that *$u^{(2m)}(x)$ be bounded in* $[0, +\infty)$ [1]) and putting $M_{2m} = \sup\limits_{x \geqq 0} |u^{(2m)}(x)|$, (4.9.9) yields by theor. 4.9.II:

$$|R(u)| \leqq \frac{m!\, \Gamma(\alpha + m + 1)}{(2m)!} M_{2m} . \tag{4.9.13}$$

[1]) The consequence of this hypothesis is that (4.9.3) is satisfied.

4.10. Gauss-Hermite formulae. These formulae are relative to the interval $(-\infty, +\infty)$ and to the weight $g(x) = e^{-x^2}$ and have the following form:

$$\boxed{\int_{-\infty}^{+\infty} e^{-x^2} u(x)\, dx = \sum_{i=1}^{m} H_{mi}\, u(x_{mi}) + R(u)}\,, \tag{4.10.1}$$

with m nodes $x_{m1}, \ldots, x_{mm}$, under the condition that $R(u) = 0$ *when* $u(x)$ *is a polynomial of degree* $\leqq 2m - 1$ $\left(\textit{that is } E = \frac{d^{2m}}{dx^{2m}}\right)$. We shall obtain (4.10.1) under hypotheses of the type (2.3.8). Operating as in § 4.9 we see that the needed hypotheses are

$$e^{-x^2} \frac{x^{i-1}}{(i-1)!} \in L(-\infty, +\infty);$$

$$(-1)^{2m-i} \frac{x^{2m-i}}{(2m-i)!} u^{(2m)}(x) \int_{-\infty}^{x} e^{-\xi^2} \frac{\xi^{i-1}}{(i-1)!}\, d\xi \in L(-\infty, 0]\,;$$

$$(-1)^{2m-i} \frac{x^{2m-i}}{(2m-1)!} u^{(2m)}(x) \int_{x}^{+\infty} e^{-\xi^2} \frac{\xi^{i-1}}{(i-i)!}\, d\xi \in L[0, +\infty)\,,$$

$$(i = 1, 2, \ldots, 2m)\,.$$

The first hypotheses are surely satisfied, while the others, considering that

$$\int_{-\infty}^{x} e^{-\xi^2} \xi^{i-1}\, d\xi \sim -\frac{1}{2} x^{i-2} e^{-x^2} \qquad (\text{for } x \to -\infty)\,;$$

$$\int_{x}^{+\infty} e^{-\xi^2} \xi^{i-1}\, d\xi \sim \frac{1}{2} x^{i-2} e^{-x^2} \qquad (\text{for } x \to +\infty)\,,$$

reduce to the sole condition

$$x^{2m-2} e^{-x^2} u^{(2m)}(x) \in L(-\infty, +\infty)\,. \tag{4.10.2}$$

Now we see how to choose the nodes in order that the Gauss problem expressed by (4.10.1) have solutions. Operating as in the preceding paragraph we immediately find that it must be

$$\int_{-\infty}^{+\infty} e^{-x^2} x^r \prod_{i=1}^{m} (x - x_{mi})\, dx = 0\,, \qquad (r = 0, 1, \ldots, m-1)\,,$$

so that (see § 3.7) *the nodes* $x_{m1}, \ldots, x_{mm}$ *must coincide with the zeros of the Hermite polynomial* $H_m(x)$ (which are set simmetrically with respect to the point $x = 0$). Choosing the nodes in this way the formula (4.10.1) is unique and if we put in it

$$u(x) = \frac{1}{H_m'(x_{mj})} \frac{H_m(x)}{x - x_{mj}}$$

we obtain

$$H_{mj} = \frac{1}{H'_m(x_{mj})} \int_{-\infty}^{+\infty} e^{-x^2} \frac{H_m(x)}{x - x_{mj}} dx\,, \quad (j = 1, 2, \ldots, m)\,. \quad (4.10.3)$$

In order to compute this integral, we examine again the Christoffel-Darboux summation formula (3.7.8) and put in it $y = x_{mj}$; in this way we obtain

$$\sum_{k=0}^{m-1} \frac{1}{2^k\, k!} H_k(x)\, H_k(x_{mj}) = -\frac{1}{2^{m+1}\, m!} \frac{H_{m+1}(x_{mj})\, H_m(x)}{x - x_{mj}}$$

from which we deduce

$$\sqrt{\pi} = -\frac{H_{m+1}(x_{mj})}{2^{m+1}\, m!} \int_{-\infty}^{+\infty} e^{-x^2} \frac{H_m(x)}{x - x_{mj}} dx\,, \quad (4.10.4)$$

by multiplying both members by e^{-x^2} and integrating from $-\infty$ to $+\infty$.

Then by virtue of (3.7.2) we have

$$H_{m+1}(x_{mj}) = -H'_m(x_{mj}) \quad (4.10.5)$$

and from (4.10.3), (4.10.4) and (4.10.5) we finally have

$$H_{mj} = \frac{\sqrt{\pi}\, 2^{m+1}\, m!}{[H'_m(x_{mj})]^2}\,, \quad (H_{mj} = H_{m,\, m+1-j})\,. \quad (4.10.6)$$

In order to apply (4.1.3) to obtain the expression of the remainder $R(u)$ we must get the solutions $\varphi_0(x), \varphi_1(x), \ldots, \varphi_m(x)$ of the equation $\varphi^{(2m)}(x) = e^{-x^2}$. It suffices to apply theor. 2.4.I to obtain

$$\varphi_i(x) = \int_{-\infty}^{x} \frac{(x-\xi)^{2m-1}}{(2\,m-1)!} e^{-\xi^2} d\xi - \sum_{j=1}^{i} H_{mj} \frac{(x - x_{mj})^{2m-1}}{(2\,m-1)!}\,,$$

$$(i = 0, 1, \ldots, m)\,,\ ^{1)} \quad (4.10.7)$$

after which we have

$$R(u) = \int_{-\infty}^{+\infty} \Phi(x)\, u^{(2m)}(x)\, dx = \sum_{i=0}^{m} \int_{x_{mi}}^{x_{m,i+1}} \varphi_i(x)\, u^{(2m)}(x)\, dx\,,$$

$$(x_{m0} = -\infty\,,\ x_{m,m+1} = +\infty)\,, \quad (4.10.8)$$

where the influence-function $\Phi(x)$ is given by

$$\Phi(x) = \varphi_i(x) \quad \text{for} \quad x_{mi} < x < x_{m,i+1}\,, \quad (i = 0, 1, \ldots, m)\,. \quad (4.10.9)$$

Theorems analogous to those of the preceding paragraph hold here and are

[1]) For $i = m$ we may also write $\varphi_m(x) = -\int_{x}^{+\infty} \frac{(x-\xi)^{2m-1}}{(2\,m-1)!} e^{-\xi^2} d\xi$.

expressed by the following formulae

$$\Phi(x) \in C^{2m-2}(-\infty, +\infty), \quad \Phi(x) > 0, \tag{4.10.10}$$

$$\int_{-\infty}^{+\infty} \Phi(x)\, dx = \frac{\sqrt{\pi}\, m!}{2^m (2m)!}, [1] \tag{4.10.11}$$

for which proofs analogous to those of the preceding paragraph are valid and whose developments we leave to the reader.

From (4.10.8), (4.10.10) there follows that *if* $u^{(2m)}(x)$ *has constant sign in* $(-\infty, +\infty)$, $R(u)$ *has the same sign*, while from (4.10.8), (4.10.11) there follows *under the hypothesis that* $u^{(2m)}(x)$ *be bounded* [and therefore satisfying (4.10.2)]:

$$|R(u)| \leqq \frac{\sqrt{\pi}\, m!}{2^m (2m)!} M_{2m} \quad \text{with} \quad M_{2m} = \sup_{-\infty < x < +\infty} |u^{(2m)}(x)|. \tag{4.10.12}$$

4.11. Newton-Cotes formulae. These formulae refer to a finite interval (which we shall suppose coinciding with $[-1, 1]$) and to the weight $g(x) = 1$. They have the following form

$$\boxed{\int_{-1}^{1} u(x)\, dx = \sum_{i=1}^{m} A_{mi}\, u(x_{mi}) + R(u)}, \tag{4.11.1}$$

and the nodes $x_{m1}, \ldots, x_{mm}$ *are set on the* m *points which divide* $[-1, 1]$ *into* $m - 1$ *equal intervals:*

$$x_{mi} = -1 + (i-1) h_m \quad \text{with} \quad h_m = \frac{2}{m-1}, \quad (i = 1, 2, \ldots, m), \tag{4.11.2}$$

under the condition that $R(u) = 0$ *when* $u(x)$ *is a polynomial of degree* $m - 1$ *(if* m *is even), of degree* m *(if* m *is odd).* Namely we have

$$E = \frac{d^m}{dx^m} \text{ (if } m \text{ is even) or } E = \frac{d^{m+1}}{dx^{m+1}} \text{ (if } m \text{ is odd)}, \tag{4.11.3}$$

while we must assume respectively $u(x) \in A\,C^{m-1}\,[-1, 1]$ or $u(x) \in A\,C^m\,[-1, 1]$. Note that the nodes x_{mi} are set symmetrically with respect to the origin; that is we have

$$x_{mi} = -x_{m, m+1-i}. \tag{4.11.4}$$

Formula (4.11.1) expresses a Gauss problem (§ 2.5) whose corresponding boundary problem is to be written

$$\frac{d^m u}{dx^m} = 0; \quad u(x_{mi}) = 0. \quad (i = 1, 2, \ldots, m) \quad (\text{if } m \text{ is even}), \tag{4.11.5}$$

$$\frac{d^{m+1} u}{dx^{m+1}} = 0; \quad u(x_{mi}) = 0, \quad (i = 1, 2, \ldots, m) \quad (\text{if } m \text{ is odd}). \tag{4.11.6}$$

[1]) Keep in mind that (see § 3.7)

$$\int_{-\infty}^{+\infty} e^{-x^2} H_m^2(x)\, dx = \sqrt{\pi}\, 2^m m! \quad \text{and} \quad H_m(x) = 2^m x^m + \cdots.$$

It is obvious that the problem (4.11.5) has only the solutions $u(x) \equiv 0$. On the contrary problem (4.11.6) have the solution $\prod_{i=1}^{m} (x - x_{mi})$, which leads to the condition $\int_{-1}^{1} \prod_{i=1}^{m} (x - x_{mi})\, dx = 0$; but this is surely satisfied since by virtue of (4.11.4) the function $\prod_{i=1}^{m} (x - x_{mi})$ is an odd function. Therefore in both cases (4.11.1) is possible; indeed it is possible only in one way since with the notations of § 2.5 we have $m(n - p) - n + q = m \cdot 1 - m + 0 = 0$ (if m is even), $= m \cdot 1 - (m + 1) + 1 = 0$ (if m is odd).

In the following it is useful to introduce the polynomial of degree m

$$\omega_m(x) = \prod_{i=1}^{m} (x - x_{mi}), \tag{4.11.7}$$

which is an even function (if m is even), odd (if m is odd). If we put in (4.11.1)

$$u(x) = \frac{1}{\omega'_m(x_{mj})} \frac{\omega_m(x)}{x - x_{mj}} \quad \text{(polynomial of degree } m - 1\text{)},$$

we immediately find

$$A_{mj} = \frac{1}{\omega'_m(x_{mj})} \int_{-1}^{1} \frac{\omega_m(x)}{x - x_{mj}}\, dx, \quad (j = 1, 2, \ldots, m), \text{[1]} \tag{4.11.8}$$

and therefore, as it is easy to check on the basis of (4.11.4):

$$A_{mj} = A_{m, m+1-j}. \tag{4.11.9}$$

To obtain the expression of the remainder $R(u)$ we must get the solutions $\varphi_1(x), \ldots, \varphi_m(x)$ of the differential equation $E^*(\varphi) = g$ which now is written

$$\varphi^{(m)}(x) = 1 \quad \text{(if } m \text{ is even)}; \qquad \varphi^{(m+1)}(x) = 1 \quad \text{(if } m \text{ is odd)}.$$

These solutions are immediately given by theorem 2.4.I; we have precisely

$$\varphi_i(x) = \frac{(x + 1)^m}{m!} - \sum_{j=1}^{i} A_{mj} \frac{(x - x_{mj})^{m-1}}{(m - 1)!}, \quad \text{(if } m \text{ is even)}, \tag{4.11.10}$$

$$\varphi_i(x) = \frac{(x + 1)^{m+1}}{(m + 1)!} - \sum_{j=1}^{i} A_{mj} \frac{(x - x_{mj})^m}{m!}, \quad \text{(if } m \text{ is odd)}, \tag{4.11.11}$$

$$(i = 1, 2, \ldots, m - 1).$$

After that, in virtue of (4.1.3), we have

$$R(u) = \int_{-1}^{1} \Phi(x)\, u^{(m)}(x)\, dx = \sum_{i=1}^{m-1} \int_{x_{mi}}^{x_{m,i+1}} \varphi_i(x)\, u^{(m)}(x)\, dx, \quad \text{(if } m \text{ is even)}, \tag{4.11.12}$$

$$R(u) = \int_{-1}^{1} \Phi(x)\, u^{(m+1)}(x)\, dx = \sum_{i=1}^{m-1} \int_{x_{mi}}^{x_{m,i+1}} \varphi_i(x)\, u^{(m+1)}(x)\, dx, \quad \text{(if } m \text{ is odd)}, \tag{4.11.13}$$

[1]) For another expression (without integral) of A_{mj} see § 4.15, Problem 6.

where the influence-function $\Phi(x)$ is expressed by

$$\Phi(x) = \varphi_i(x) \quad \text{for} \quad x_{mi} < x < x_{m,i+1}, \qquad (i = 1, 2, \ldots, m-1). \qquad (4.11.14)$$

To deduce from the last formulae a suitable evaluation of the remainder $R(u)$, we begin by assuming that

$$u(x) \in C^m[-1, 1] \quad (\text{if } m \text{ is even}), \qquad u(x) \in C^{m+1}[-1, 1] \quad (\text{if } m \text{ is odd}) \qquad (4.11.15)$$

and to get, under such hypotheses, another expression for $R(u)$. To this purpose we introduce the following function

$$U_m(x) = \omega_m(x)\, u(x_{m1}, x_{m2}, \ldots, x_{mm}, x) \qquad (4.11.16)$$

where $\omega_m(x)$ is a polynomial determined by (4.11.7) and $u(x_{m1}, x_{m2}, \ldots, \ldots, x_{mm}, x)$ is the function introduced in § 3.8 (divided difference of m-th order) for which, as a consequence of (4.11.15), we have

$$\left.\begin{aligned} &u(x_{m1}, \ldots, x_{mm}, x) \in C^{m-1}[-1, 1] \quad &(\text{if } m \text{ is even});\\ &u(x_{m1}, \ldots, x_{mm}, x) \in C^{m}[-1, 1] \quad &(\text{if } m \text{ is odd}). \end{aligned}\right\} \qquad (4.11.17)$$

For (3.8.1) we may write

$$U_m(x) = \frac{\omega_m(x)}{V(x_{m1}, \ldots, x_{mm}, x)} \begin{vmatrix} 1 & \ldots & 1 & 1 \\ x_{m1} & \ldots & x_{mm} & x \\ x_{m1}^2 & \ldots & x_{mm}^2 & x^2 \\ \cdot & \cdot & \cdot & \cdot \\ x_{m1}^{m-1} & \ldots & x_{mm}^{m-1} & x^{m-1} \\ u(x_{m1}) & \ldots & u(x_{mm}) & u(x) \end{vmatrix}$$

$$= \frac{1}{V(x_{m1}, \ldots, x_{mm})} \begin{vmatrix} 1 & \ldots & 1 & 1 \\ x_{m1} & \ldots & x_{mm} & x \\ x_{m1}^2 & \ldots & x_{mm}^2 & x^2 \\ \cdot & \cdot & \cdot & \cdot \\ x_{m1}^{m-1} & \ldots & x_{mm}^{m-1} & x^{m-1} \\ u(x_{m1}) & \ldots & u(x_{mm}) & u(x) \end{vmatrix}$$

from which we have $U_m^{(m)}(x) = u^{(m)}(x)$ (if m is even), $U_m^{(m+1)}(x) = u^{(m+1)}(x)$ (if m is odd). Therefore from (4.11.12), (4.11.13) there follows $R(u) = R(U_m)$; on the other part from (4.11.16), (4.11.17) we get $U_m(x_{mi}) = 0, (i = 1, 2, \ldots, m)$, so that (4.11.1) yields $\int_{-1}^{1} U_m(x)\,dx = R(U_m)$. After all, we may write $R(u) = \int_{-1}^{1} U_m(x)\,dx$ that is

$$R(u) = \int_{-1}^{1} \omega_m(x)\, u(x_{m1}, \ldots, x_{mm}, x)\,dx\,. \qquad (4.11.18)$$

This is the form of $R(u)$ which better allows its evaluation, and for obtaining it we need to state some simple theorems.

Theorem 4.11.I. *For* $-1 - h_m < x \leqq -h_m$ *we have*

$$|\omega_m(x + h_m)| < |\omega_m(x)| . \qquad (4.11.19)$$

Proof. Recalling (4.11.7) we may then write

$$\left|\frac{\omega_m(x + h_m)}{\omega_m(x)}\right| = \left|\frac{x + h_m + 1}{x - 1}\right| = \frac{x + h_m + 1}{1 - x} \leqq \frac{1}{1 + h_m} < 1 .$$

Theorem 4.11.II. *If m is odd, putting*

$$\Omega_m(x) = \int_{-1}^{x} \omega_m(t)\, dt , \qquad (4.11.20)$$

we have

$$\Omega_m(-1) = \Omega_m(1) = 0 , \qquad (4.11.21)$$

$$\Omega_m(x) > 0 \quad \text{for} \quad -1 < x < 1 , \qquad (4.11.22)$$

$$\int_{-1}^{1} x\, \omega_m(x)\, dx < 0 . \qquad (4.11.23)$$

Proof. Formulae (4.11.21) are clear, considering that $\omega_m(x)$ is an odd function. As far as (4.11.22) is concerned it suffices to prove it for $-1 < x \leqq 0$, since $\Omega_m(x)$ is obviously an even function. Since $\omega_m(x) > 0$ for $x_{m1} < x < x_{m2}$, $\omega_m(x) < 0$ for $x_{m2} < x < x_{m3}, \ldots$ (see Fig. 5), to prove the thesis it is

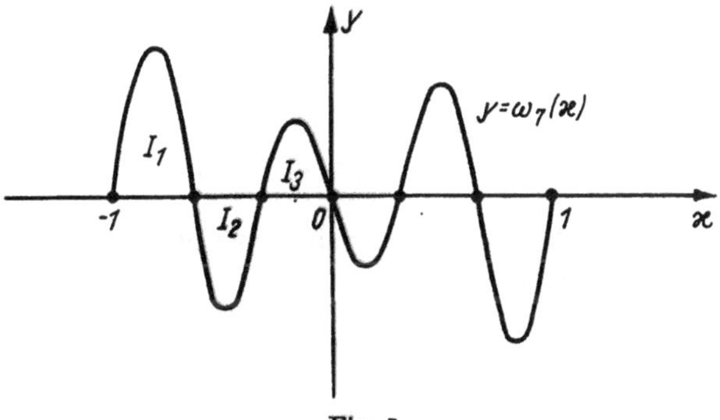

Fig. 5

sufficient show that, putting $I_s = \int_{x_{ms}}^{x_{m,s+1}} \omega_m(x)\, dx$, $\left(s = 1, 2, \ldots, \frac{m-1}{2}\right)$, we have $|I_{s+1}| < |I_s|$, $\left(s = 1, 2, \ldots, \frac{m-3}{2}\right)$. In fact we have

$$|I_{s+1}| = \left|\int_{x_{m,s+1}}^{x_{m,s+2}} \omega_m(x)\, dx\right| = \int_{x_{m,s+1}}^{x_{m,s+2}} |\omega_m(x)|\, dx = \int_{x_{m,s}}^{x_{m,s+1}} |\omega_m(\xi + h_m)|\, d\xi ;$$

but $x_{ms} \leqq \xi \leqq x_{m,s+1}$ $\left(s = 1, 2, \ldots, \frac{m-3}{2}\right)$ involves $-1 - h_m < \xi \leqq -h_m$

and therefore, on the basis of the preceding theorem, in the last integral we have $|\omega_m(\xi + h_m)| < |\omega_m(\xi)|$.

There follows

$$|I_{s+1}| < \int_{x_{ms}}^{x_{m,s+1}} |\omega_m(\xi)| \, d\xi = |I_s|$$

and then (4.11.22) is proved. From it we get $\int_{-1}^{1} \Omega_m(x)\, dx > 0$ but, using (4.11.20) and (4.11.21), we may write

$$\int_{-1}^{1} \Omega_m(x)\, dx = [x\, \Omega_m(x)]_{-1}^{1} - \int_{-1}^{1} x\, \Omega_m'(x)\, dx = - \int_{-1}^{1} x\, \omega_m(x)\, dx \quad (4.11.24)$$

and there follows (4.11.23), q.e.d.

Obviously this theorem holds if we substitute the interval $[-1, 1]$ by any interval $[a, b]$ and we divide it into an even number $m - 1$ of equal parts. On account of this fact we may state the following theorem:

Theorem 4.11.III. *If m is even, putting*

$$\left.\begin{aligned} &\omega_{m-1}^*(x) = (x - x_{m1})(x - x_{m2}) \ldots (x - x_{m,m-1}) \text{ and therefore} \\ &\omega_m(x) = (x - x_{mm})\, \omega_{m-1}^*(x) , \end{aligned}\right\} \quad (4.11.25)$$

$$\Omega_{m-1}^*(x) = \int_{-1}^{x} \omega_{m-1}^*(t)\, dt , \quad (4.11.26)$$

we have

$$\Omega_{m-1}^*(-1) = \Omega_{m-1}^*(x_{m,m-1}) = 0 , \quad (4.11.27)$$

$$\Omega_{m-1}^*(x) > 0 \quad \text{for } -1 < x < x_{m,m-1} , \quad (4.11.28)$$

$$\int_{-1}^{x_{m,m-1}} x\, \omega_{m-1}^*(x)\, dx < 0 . \quad (4.11.29)$$

We may now deduce the requested evaluations of the remainder $R(u)$, by means of the two following theorems:

Theorem 4.11.IV. *If m is odd and $u(x) \in C^{m+1}[-1, 1]$, we have*

$$R(u) = \frac{u^{(m+1)}(\xi)}{(m+1)!} \int_{-1}^{1} x\, \omega_m(x)\, dx , \quad \left[\text{with} \int_{-1}^{1} x\, \omega_m(x)\, dx < 0\right], \quad (4.11.30)$$

where ξ denotes a suitable point of the interval $[-1, 1]$.

Proof. Starting from (4.11.18), integrating by parts and using (4.11.20) we get

$$R(u) = [u(x_{m1}, \ldots, x_{mm}, x)\, \Omega_m(x)]_{-1}^{1} - \int_{-1}^{1} \Omega_m(x) \frac{d}{dx} u(x_{m1}, \ldots, x_{mm}, x)\, dx .$$

The first term of the second member is zero in virtue of (4.11.21); in the second term the factor $\Omega_m(x)$ has constant sign [for (4.11.22)] and therefore we may apply to the integral the middle value theorem; we then obtain

$$R(u) = -\left[\frac{d}{dx}u(x_{m1}, \ldots, x_{mm}, x)\right]_{x=\bar{x}} \int_{-1}^{1} \Omega_m(x)\,dx\,, \qquad (\text{with } \bar{x} \in [-1, 1])\,,$$

and applying theorem 3.8.II

$$R(u) = -\frac{u^{(m+1)}(\xi)}{(m+1)!} \int_{-1}^{1} \Omega_m(x)\,dx\,.$$

Whence, using (4.11.23) and (4.11.24), we soon get (4.11.30), q.e.d.

Theorem 4.11.V. *If m is even and $u(x) \in C^m[-1, 1]$ we have*

$$R(u) = \frac{u^{(m)}(\xi)}{m!} \int_{-1}^{1} \omega_m(x)\,dx\,, \quad \left[\text{with} \int_{-1}^{1} \omega_m(x)\,dx < 0\right], \tag{4.11.31}$$

where ξ denotes a suitable point of the interval $[-1, 1]$.

Proof. Starting from (4.11.18) we write

$$R(u) = I_1(u) + I_2(u) \tag{4.11.32}$$

with

$$I_1(u) = \int_{-1}^{x_{m,m-1}} \omega_m(x)\,u(x_{m1}, \ldots, x_{mm}, x)\,dx\,,$$

$$I_2(u) = \int_{x_{m,m-1}}^{1} \omega_m(x)\,u(x_{m1}, \ldots, x_{mm}, x)\,dx\,.$$

For $x_{m,m-1} < x < 1$ we have $\omega_m(x) < 0$ and therefore we may apply the middle value theorem to the integral $I_2(u)$, thus getting

$$I_2(u) = u(x_{m1}, \ldots, x_{mm}, \bar{x}) \int_{x_{m,m-1}}^{1} \omega_m(x)\,dx\,, \qquad (\text{with } \bar{x} \in [x_{m,m-1}, 1])$$

and successively, for theor. 3.8.I

$$I_2(u) = \frac{u^{(m)}(\xi_2)}{m!} \int_{x_{m,m-1}}^{1} \omega_m(x)\,dx\,, \quad \left(\text{with } \xi_2 \in [-1, 1] \text{ and} \int_{x_{m,m-1}}^{1} \omega_m(x)\,dx < 0\right). \tag{4.11.33}$$

As to $I_1(u)$, using (4.11.25) and (4.11.26) and applying the formula (3.8.3) we may write

$$I_1(u) = \int_{-1}^{x_{m,m-1}} (x - x_{mm})\,\omega^*_{m-1}(x)$$

$$\times \frac{u(x_{m1}, \ldots, x_{m,m-1}, x) - u(x_{m1}, \ldots, x_{m,m-1}, x_{mm})}{x - x_{mm}}\,dx$$

$$= \int_{-1}^{x_{m,m-1}} \omega^*_{m-1}(x)\,u(x_{m1}, \ldots, x_{m,m-1}, x)\,dx$$

$$- u(x_{m1}, \ldots, x_{m,m-1}, x_{mm})\,\Omega^*_{m-1}(x_{m,m-1})$$

and therefore, for (4.11.27):

$$I_1(u) = \int_{-1}^{x_{m,m-1}} \omega_{m-1}^*(x)\, u(x_{m1}, \dots, x_{m,m-1}, x)\, dx\,.$$

On this integral, considering that $m - 1$ is odd, we may reason analogously as in the proof of the preceding theorem and deduce

$$I_1(u) = \frac{u^{(m)}(\xi_1)}{m!} \int_{-1}^{x_{m,m-1}} x\, \omega_{m-1}^*(x)\, dx\,,$$

$$\left(\text{with } \xi_1 \in [-1, x_{m,m-1}] \quad \text{and} \quad \int_{-1}^{x_{m,m-1}} x\, \omega_{m-1}^*(x)\, dx < 0\right).$$

But, using (4.11.25), (4.11.26) and (4.11.27), we may write

$$\begin{aligned} \int_{-1}^{x_{m,m-1}} x\, \omega_{m-1}^*(x)\, dx &= \int_{-1}^{x_{m,m-1}} (x - x_{mm})\, \omega_{m-1}^*(x)\, dx + x_{mm}\, \Omega_{m-1}^*(x_{m,m-1}) \\ &= \int_{-1}^{x_{m,m-1}} \omega_m(x)\, dx \end{aligned}$$

and therefore

$$I_1(u) = \frac{u^{(m)}(\xi_1)}{m!} \int_{-1}^{x_{m,m-1}} \omega_m(x)\, dx\,,$$

$$\left(\text{with } \xi_1 \in [-1, x_{m,m-1}] \quad \text{and} \quad \int_{-1}^{x_{m,m-1}} \omega_m(x)\, dx < 0\right). \tag{4.11.34}$$

From (4.11.32), (4.11.33) and (4.11.34) there follows

$$R(u) = \frac{1}{m!}[p\, u^{(m)}(\xi_1) + q\, u^{(m)}(\xi_2)] \int_{-1}^{1} \omega_m(x)\, dx\,, \quad \left[\text{with} \int_{-1}^{1} \omega_m(x)\, dx < 0\right], \tag{4.11.35}$$

where we have introduced the two *positive* numbers

$$p = \frac{\int_{-1}^{x_{m,m-1}} \omega_m(x)\, dx}{\int_{-1}^{1} \omega_m(x)\, dx}\,, \qquad q = \frac{\int_{x_{m,m-1}}^{1} \omega_m(x)\, dx}{\int_{-1}^{1} \omega_m(x)\, dx}$$

whose sum is 1. Then it is evident that $p\, u^{(m)}(\xi_1) + q\, u^{(m)}(\xi_2)$ is a value taken by the continuous function $u(x)$ in $[-1, 1]$ and therefore (4.11.35) soon becomes (4.11.31) q.e.d.

From the two last theorems we deduce these others:

Theorem 4.11.VI. *With any m, the influence-function $\Phi(x)$ is negative or null in* $[-1, 1]$:

$$\Phi(x) \leqq 0\,, \qquad (-1 \leqq x \leqq 1)\,, \tag{4.11.36}$$

and moreover we have

$$\int_{-1}^{1} \Phi(x)\, dx = \begin{cases} \dfrac{1}{(m+1)!} \displaystyle\int_{-1}^{1} x\, \omega_m(x)\, dx\,, & (\textit{if } m \textit{ is odd})\,, \\ \dfrac{1}{m!} \displaystyle\int_{-1}^{1} \omega_m(x)\, dx\,, & (\textit{if } m \textit{ is even})\,. \end{cases} \tag{4.11.37}$$

Proof. Now we prove (4.11.36) assuming, for instance, that m be odd. From (4.11.30) there follows that, if we have $u^{(m+1)}(x) \geqq 0$ in $[-1, 1]$, certainly there results $R(u) \leqq 0$. Whence, if $\Phi(x) > 0$ should be in any interval contained in $[-1, 1]$, we could obviously build up a function $u(x) \in C^{m+1}$ $[-1, 1]$ such as to give

$$u^{(m+1)}(x) = 0 \quad \text{where} \quad \Phi(x) \leqq 0\,, \qquad u^{(m+1)}(x) > 0 \quad \text{where} \quad \Phi(x) > 0\,.$$

For such $u(x)$ we should have, on the basis of the preceding remark, $R(u) \leqq 0$, while, for the way in which $u(x)$ was built up, (4.11.13) should yield evidently $R(u) > 0$. This proves that (4.11.36) holds.

To establish (4.11.37) is suffices to remark that from (4.11.12), (4.11.13), (4.11.30) and (4.11.31) there follow

$$\int_{-1}^{1} \Phi(x)\, u^{(m+1)}(x)\, dx = \frac{u^{(m+1)}(\xi)}{(m+1)!} \int_{-1}^{1} x\, \omega_m(x)\, dx \qquad (\text{if } m \text{ is odd})\,,$$

$$\int_{-1}^{1} \Phi(x)\, u^{(m)}(x)\, dx = \frac{u^{(m)}(\xi)}{m!} \int_{-1}^{1} \omega_m(x)\, dx \qquad (\text{if } m \text{ is even})\,,$$

and to put in them respectively $u(x) = x\, \omega_m(x)$, $u(x) = \omega_m(x)$.

Theorem 4.11.VII. *Under the hypothesis that $u^{(m+1)}(x)$ (if m is odd) or $u^{(m)}(x)$ (if m is even) be bounded in $[-1, 1]$, we have*

$$\left.\begin{aligned} |R(u)| &\leqq \frac{M_{m+1}}{(m+1)!} \left| \int_{-1}^{1} x\, \omega_m(x)\, dx \right| && (\text{if } m \text{ is odd})\;; \\ |R(u)| &\leqq \frac{M_m}{m!} \left| \int_{-1}^{1} \omega_m(x)\, dx \right| && (\text{if } m \text{ is even})\,, \end{aligned}\right\} \tag{4.11.38}$$

where M_{m+1} [or M_m] *denotes the lower upper bound of* $|u^{(m+1)}(x)|$ [or $|u^{(m)}(x)|$] *in* $[-1, 1]$.

Proof. For instance, in the case that m is odd, we have for (4.11.13), (4.11.36) and (4.11.37):

$$|R(u)| \leqq \int_{-1}^{1} |\Phi(x)|\,|u^{(m+1)}(x)|\,dx \leqq M_{m+1} \int_{-1}^{1} |\Phi(x)|\,dx$$

$$= M_{m+1} \left| \int_{-1}^{1} \Phi(x)\,dx \right| = \frac{M_{m+1}}{(m+1)!} \left| \int_{-1}^{1} x\,\omega_m(x)\,dx \right| .$$

The computation of the integrals occurring in (4.11.30), (4.11.31), (4.11.38) is easily carried out for any preassigned value of m (see § 4.15, Problem 9).

In the cases $m = 2$, $m = 3$ the formula (4.11.1) reduces to the formula of the inscribed trapezium [see (4.2.6)] and to the Cavalieri-Simpson formula [see (4.2.11)].

4.12. Examples of Tchebychef problems. As first example, we shall build up a formula of the type

$$\int_{-1}^{1} \frac{u(x)}{\sqrt{1-x^2}}\,dx = A \sum_{j=1}^{m} u(x_j) + R(u) \,, \tag{4.12.1}$$

with $R(u) = 0$ when $u(x)$ is a polynomial of degree $\leqq 2m - 1$ $\left(\text{that is } E = \frac{d^{2m}}{dx^{2m}}\right)$. According to theor. 2.6.I we must consider the boundary problem $u^{(2m)}(x) = 0$, $\sum_{j=1}^{m} u(x_j) = 0$, which admits the following $2m - 1$ linearly independent solutions $U_k(x) = x^k - \frac{s_k}{m}$, $(k = 1, 2, \ldots, 2m - 1)$, having put

$$s_k = \sum_{j=1}^{m} x_j^k ; \tag{4.12.2}$$

there follows that (4.12.1) is possible only if

$$\int_{-1}^{1} \left(x^k - \frac{s_k}{m}\right) \frac{dx}{\sqrt{1-x^2}} = 0 \,, \qquad (k = 1, 2, \ldots, 2m - 1) \,,$$

namely, operating the substitution $x = \cos t$:

$$\int_{0}^{\pi} \left(\cos^k t - \frac{s_k}{m}\right) dt = 0 \,, \qquad (k = 1, 2, \ldots, 2m - 1) \,.$$

The nodes $x_1, \ldots, x_m$ must therefore be selected in such a way as to give

$$s_k = \frac{m}{\pi} \int_{0}^{\pi} \cos^k t\,dt = \begin{cases} 0 \,, & (k = 1, 3, \ldots, 2m - 1) \,, \\ m\,\dfrac{(k-1)!!}{k!!} \,, & (k = 2, 4, \ldots, 2m - 2) \,. \end{cases} \tag{4.12.3}$$

As it is known, these conditions (indeed only those with $k = 1, 2, \ldots, m$) determine uniquely the nodes $x_1, \ldots, x_m$ and the fact that $s_k = 0$ for k odd, means that they will be symmetric with respect to the origin; that is we shall have

$$x_{m-j+1} = -x_j\,, \qquad (j = 1, 2, \ldots, m)\,. \tag{4.12.4}$$

We check that (4.12.3) are satisfied by

$$x_j = \cos\frac{(2j-1)\pi}{2m}\,, \qquad (j = 1, 2, \ldots, m)\,, \tag{4.12.5}$$

which are acceptable values since are distinct and included between -1 and 1.

That $s_k = 0$ for k odd be satisfied is evident from the fact that (4.12.4) hold; we have only to check the formula

$$s_{2r} = \sum_{j=1}^{m} \cos^{2r}\frac{(2j-1)\pi}{2m} = m\,\frac{(2r-1)!!}{(2r)!!}\,, \quad (r = 1, 2, \ldots, m-1)\,. \tag{4.12.6}$$

First of all we may write

$$s_{2r} = \frac{1}{2^{2r}}\sum_{j=1}^{m}\left(e^{i\frac{(2j-1)\pi}{2m}} + e^{-i\frac{(2j-1)\pi}{2m}}\right)^{2r}$$

$$= \frac{1}{2^{2r}}\sum_{j=1}^{m}\sum_{s=0}^{2r}\binom{2r}{s} e^{is\frac{(2j-1)\pi}{2m}} e^{-i(2r-s)\frac{(2j-1)\pi}{2m}} = \frac{1}{2^{2r}}\sum_{s=0}^{2r}\binom{2r}{s}\sum_{j=1}^{m} e^{i(s-r)\frac{(2j-1)\pi}{m}}\,.$$

The sum with respect to the index j occurring here is a sum of m terms in geometric progression with ratio $e^{i(s-r)\frac{2\pi}{m}}$ and this can have value 1 only for $s = r$.[1] Therefore

$$\sum_{j=1}^{m} e^{i(s-r)\frac{(2j-1)\pi}{m}} = \begin{cases} e^{i(s-r)\frac{\pi}{m}}\dfrac{1 - e^{i(s-r)2\pi}}{1 - e^{i(s-r)\frac{2\pi}{m}}} = 0\,, & (\text{if } s \neq r)\,, \\[2ex] m\,, & (\text{if } s = r)\,, \end{cases}$$

whence we conclude that

$$s_{2r} = \frac{1}{2^{2r}}\binom{2r}{r} m\,.$$

From here we easily pass to (4.12.6) and thus it is established that the nodes x_j must be those given by (4.12.5). Putting then $u(x) = 1$ in (4.12.1) we soon find $A = \dfrac{\pi}{m}$ and it is clear that *(4.12.1) coincides with (4.6.25)*, already obtained as a particular case of the Gauss-Jacobi formulae.

We examine this other example:

$$\int_{-1}^{1} u(x)\,dx = A\sum_{i=1}^{m} u(x_i) + R(u)\,, \tag{4.12.7}$$

[1]) In fact $s - r$ varies between $-r$ and r (with $r \leq m - 1$) and the only multiple of m that it meets is zero.

under the condition that $R(u) = 0$ when $u(x)$ is a polynomial of degree $\leqq m + 1$ (if m is even), of degree $\leqq m$ (if m is odd); that is we assume that $E = \frac{d^{m+2}}{dx^{m+2}}$ (if m is even), $E = \frac{d^{m+1}}{dx^{m+1}}$ (if m is odd).

According to theorem 2.6.I we must consider the following boundary problems:

$$u^{(m+2)}(x) = 0\,, \qquad \sum_{i=1}^{m} u(x_i) = 0\,, \qquad \text{(if } m \text{ is even)}\,,$$

$$u^{(m+1)}(x) = 0\,, \qquad \sum_{i=1}^{m} u(x_i) = 0\,, \qquad \text{(if } m \text{ is odd)}\,,$$

which have the linearly independent solutions

$$x^k - \frac{s_k}{m}\,, \qquad (k = 1, 2, \ldots, m+1)\,, \qquad \text{(if } m \text{ is even)}\,;$$

$$x^k - \frac{s_k}{m}\,, \qquad (k = 1, 2, \ldots, m)\,, \qquad \text{(if } m \text{ is odd)}\,,$$

where we have again used the positions (4.12.2). Therefore (4.12.7) is possible only if

$$\left.\begin{aligned}
&\int_{-1}^{1}\left(x^k - \frac{s_k}{m}\right)dx = 0 \quad \text{or} \quad s_k = \begin{cases} \dfrac{m}{k+1} & (k = 2, 4, \ldots, m) \\ 0 & (k = 1, 3, \ldots, m+1) \\ \text{(if } m \text{ is even)}\,, & \end{cases} \\
&\int_{-1}^{1}\left(x^k - \frac{s_k}{m}\right)dx = 0 \quad \text{or} \quad s_k = \begin{cases} \dfrac{m}{k+1} & (k = 2, 4, \ldots, m-1) \\ 0 & (k = 1, 3, \ldots, m) \\ \text{(if } m \text{ is odd)}\,. & \end{cases}
\end{aligned}\right\} \tag{4.12.8}$$

In both cases the m nodes $x_1, \ldots, x_m$ are uniquely determined by the knowledge of $s_1, s_2, \ldots, s_m$ and since $s_k = 0$ for k odd, (4.12.4) holds. In the case of m even, we have a condition one more ($s_{m+1} = 0$) but, being $m + 1$ odd, this is a consequence of the preceding ones.

We remember from algebra that, having preassigned the values $s_1, s_2, \ldots, \ldots, s_m$, the unknown nodes $x_1, x_2, \ldots, x_m$ must be roots of the following algebraic equations of degree m:

$$\begin{vmatrix}
x^m & x^{m-1} & x^{m-2} & x^{m-3} & \ldots x & 1 \\
s_1 & 1 & 0 & 0 & \ldots 0 & 0 \\
s_2 & s_1 & 2 & 0 & \ldots 0 & 0 \\
s_3 & s_2 & s_1 & 3 & \ldots 0 & 0 \\
\cdot & \cdot & \cdot & \cdot & \cdot & \cdot \\
s_m & s_{m-1} & s_{m-2} & s_{m-3} & \ldots s_1 & m
\end{vmatrix} = 0\,. \tag{4.12.9}$$

It is possible to prove that, *with the values of s_k given by (4.12.8), this equation (4.12.9) has m real roots, distinct and included between -1 and 1 only for the following values of m:*

$$m = 1, 2, 3, 4, 5, 6, 7, 9 \,. \text{ [1]} \tag{4.12.10}$$

Therefore (4.12.7) can be built up only in these 8 cases, placing the nodes in the roots of the following polynomials, easily deducible from (4.12.9) using (4.12.8):

$$\left.\begin{aligned}
&\omega_1(x) = x\,, \qquad \omega_2(x) = x^2 - \frac{1}{3}\,, \qquad \omega_3(x) = x^3 - \frac{1}{2}\,x\,,\\
&\omega_4(x) = x^4 - \frac{2}{3}\,x^2 + \frac{1}{45}\,, \qquad w_5(x) = x^5 - \frac{5}{6}\,x^3 + \frac{7}{72}\,x\,,\\
&\omega_6(x) = x^6 - x^4 + \frac{1}{5}\,x^2 - \frac{1}{105}\,, \qquad \omega_7(x) = x^7 - \frac{7}{6}\,x^5 + \frac{119}{360}\,x^3 - \frac{149}{6480}\,x\,,\\
&\omega_9(x) = x^9 - \frac{3}{2}\,x^7 + \frac{27}{40}\,x^5 - \frac{57}{560}\,x^3 + \frac{53}{22400}\,x\,.
\end{aligned}\right\} \tag{4.12.11}$$

Putting then $u(x) = 1$ in (4.12.7) we find $A = \dfrac{2}{m}$ so that we may conclude that under the hypothesis $u(x) \in A\,C^{m+1}\,[-1, 1]$ (if m is even), $u(x) \in A\,C^m\,[-1, 1]$ (if m is odd) the following formula holds

$$\boxed{\int_{-1}^{1} u(x)\,dx = \frac{2}{m}\sum_{i=1}^{m} u(x_i) + R(u)} \tag{4.12.12}$$

only for the values of m shown in (4.12.10) with the m nodes $x_1, \ldots, x_m$ placed in the roots of the polynomials $\omega_m(x)$ determined by (4.2.11); this is exact when $u(x)$ is a polynomial of degree $\leqq m + 1$ (for $m = 2, 4, 6$), of degree $\leqq m$ (for $m = 1, 3, 5, 7, 9$). As usually, the expression of the remainder $R(u)$ is given by the formulae

$$R(u) = \int_{-1}^{1} \Phi(x)\,u^{(m+2)}(x)\,dx = \sum_{i=0}^{m} \int_{x_i}^{x_{i+1}} \varphi_i(x)\,u^{(m+2)}(x)\,dx\,, \quad (m = 2, 4, 6)\,, \tag{4.12.13}$$

$$R(u) = \int_{-1}^{1} \Phi(x)\,u^{(m+1)}(x)\,dx = \sum_{i=0}^{m} \int_{x_i}^{x_{i+1}} \varphi_i(x)\,u^{(m+1)}(x)\,dx\,, \quad (m = 1, 3, 5, 7, 9)\,, \tag{4.12.14}$$

[1]) We do not give here the proof, which can be found for instance in V. I. Krylov [1], Ch. 10.

where $x_0 = -1$, $x_{m+1} = 1$ and the functions $\varphi_i(x)$, obtained according to theor. 2.4.I, are expressed by

$$\left.\begin{aligned}
\varphi_0(x) &= \frac{(x+1)^{m+2}}{(m+2)!} ; \\
\varphi_i(x) &= \frac{(x+1)^{m+2}}{(m+2)!} - \frac{2}{m}\sum_{j=1}^{i}\frac{(x-x_j)^{m+1}}{(m+1)!}, \quad (i = 1, 2, \dots, m-1) ; \\
\varphi_m(x) &= \frac{(x-1)^{m+2}}{(m+2)!}, \quad (m = 2, 4, 6) , \\
\varphi_0(x) &= \frac{(x+1)^{m+1}}{(m+1)!} ; \\
\varphi_i(x) &= \frac{(x+1)^{m+1}}{(m+1)!} - \frac{2}{m}\sum_{j=1}^{i}\frac{(x-x_j)^{m}}{m!}, \quad (i = 1, 2, \dots, m-1) ; \\
\varphi_m(x) &= \frac{(x-1)^{m+1}}{(m+1)!}, \quad (m = 1, 3, 5, 7, 9) .
\end{aligned}\right\} \qquad (4.12.15)$$

As an easy consequence of the preceding formulae and of the symmetry of the nodes with respect to the origin, the reader will be able to prove easily *that in any case the influence-function $\Phi(x)$ is an even function.*

To get now some evaluations of the remainder, we shall operate as in the preceding paragraph when dealing with the Newton-Cotes formulae, beginning by doing in $u(x)$ the more restrictive hypotheses:

$$\left.\begin{aligned}
u(x) &\in C^{m+2}[-1, 1] \quad (\text{if } m = 2, 4, 6) ; \\
u(x) &\in C^{m+1}[-1, 1] \quad (\text{if } m = 1, 3, 5, 7, 9)
\end{aligned}\right\} \qquad (4.12.16)$$

and considering this other expression of the remainder $R(u)$

$$R(u) = \int_{-1}^{1} \omega_m(x)\, u(x_1, \dots, x_m, x)\, dx \qquad (4.12.17)$$

where the polynomials (4.12.11) and a divided difference of m-th order of $u(x)$ occur.

For m even, it is convenient to introduce the following function

$$\Omega_m(x) = \int_{-1}^{x} dt \int_{-1}^{t} \omega_m(s)\, ds , \quad (m = 2, 4, 6) , \qquad (4.12.18)$$

for which from (4.12.11) it is easy to deduce

$$\left.\begin{aligned}
&\Omega_2(x) = \frac{1}{12}(x^2-1)^2 , \qquad \Omega_4(x) = \frac{1}{90}(x^2-1)^2(3x^2+1) , \\
&\Omega_6(x) = \frac{1}{840}(x^2-1)^2(15x^4+2x^2+3) ,
\end{aligned}\right\} \qquad (4.12.19)$$

whence it is evident that

$$\left.\begin{aligned}&\Omega_m(-1) = \Omega'_m(-1) = \Omega_m(1) = \Omega'_m(1) = 0\ ;\\ &\Omega_m(x) > 0\,, \quad (-1 < x < 1)\ ; \quad (m = 2, 4, 6)\,.\end{aligned}\right\} \tag{4.12.20}$$

We may now prove the following theorem:

Theorem 4.12.I. *If $m = 2, 4, 6$ and if $u(x)$ satisfies the hypothesis (4.12.16), there holds the formula*

$$R(u) = 2\frac{u^{(m+2)}(\xi)}{(m+2)!} I_m \qquad \text{with} \qquad I_m = \int_{-1}^{1} \Omega_m(x)\, dx > 0\,, \tag{4.12.21}$$

where ξ is a suitable point in $[-1, 1]$. The value of the integrals I_m are the following

$$I_2 = \frac{4}{45}\,, \qquad I_4 = \frac{16}{945}\,, \qquad I_6 = \frac{8}{1575}\,. \tag{4.12.22}$$

Proof. We start from (4.12.17) and we integrate by parts, considering that for (4.12.18) we may assume as integral of $\omega_m(x)$ the function $\Omega'_m(x) = \int_{-1}^{x} \omega_m(s)\, ds$; in this way we get

$$R(u) = [u(x_1, \ldots, x_m, x)\, \Omega'_m(x)]_{-1}^{1} - \int_{-1}^{1} \Omega'_m(x) \frac{d}{dx} u(x_1, \ldots, x_m, x)\, dx\,;$$

and therefore using (4.12.20)

$$R(u) = -\int_{-1}^{1} \Omega'_m(x) \frac{d}{dx} u(x_1, \ldots, x_m, x)\, dx\,.$$

Integrating once more by parts and using again (4.12.20) we get

$$R(u) = \int_{-1}^{1} \Omega_m(x) \frac{d^2}{dx^2} u(x_1, \ldots, x_m, x)\, dx\,.$$

Since $\Omega_m(x)$ does not change its sign in $[-1, 1]$, we may apply the middle value theorem and write down

$$R(u) = \left[\frac{d^2}{dx^2} u(x_1, \ldots, x_m, x)\right]_{x=\bar{x}} \int_{-1}^{1} \Omega_m(x)\, dx\,, \qquad (\text{with } \bar{x} \in [-1, 1])\,,$$

from which, applying theor. 3.8.III, we get (4.12.21). Formulae (4.12.22) follow soon from (4.12.19), q.e.d.

If m is odd, it is convenient instead to introduce the function

$$\Omega_m(x) = \int_{-1}^{x} \omega_m(t)\, dt\,, \qquad (m = 1, 3, 5, 7, 9) \tag{4.12.23}$$

for which we easily deduce from (4.12.11)

$$\left.\begin{aligned}
\Omega_1(x) &= \frac{1}{2}(x^2-1)\,, \quad \Omega_3(x) = \frac{1}{4}(x^2-1)\,x^2\,,\\
\Omega_5(x) &= \frac{1}{144}(x^2-1)\left[24\left(x^2-\frac{1}{8}\right)^2+\frac{5}{8}\right],\\
\Omega_7(x) &= \frac{1}{12960}(x^2-1)\left[1620\left(x^3-\frac{5}{18}x\right)^2+46\,x^2+22\right],\\
\Omega_9(x) &= \frac{1}{44800}\\
&\quad\times(x^2-1)\left[4480\left(x^4-\frac{7}{16}x^2+\frac{15}{512}\right)^2+\frac{3035}{32}x^2+\frac{59709}{2048}\right],
\end{aligned}\right\} \quad (4.12.24)$$

whence we have

$$\Omega_m(-1)=\Omega_m(1)=0\;;\quad \Omega_m(x)<0\,,\quad (-1<x<1)\;;\;(m=1,3,5,7,9)\,. \quad (4.12.25)$$

We then have the following theorem

Theorem 4.12.II. *If $m = 1, 3, 5, 7, 9$ and if $u(x)$ satisfies hypothesis (4.12.16) the following formula holds*

$$R(u)=\frac{u^{(m+1)}(\xi)}{(m+1)!}I_m \quad\text{with}\quad I_m=-\int_{-1}^{1}\Omega_m(x)\,dx>0\,, \quad (4.12.26)$$

where ξ is a suitable point in $[-1, 1]$. The values of the integrals I_m are the following

$$I_1=\frac{2}{3}\,,\quad I_3=\frac{1}{15}\,,\quad I_5=\frac{13}{756}\,,\quad I_7=\frac{281}{48600}\,,\quad I_9=\frac{163}{73920}\,. \quad (4.12.27)$$

Proof. Integrating by parts and using (4.12.23) and (4.12.25), from (4.12.17) we obtain:

$$R(u)=-\int_{-1}^{1}\Omega_m(x)\frac{d}{dx}u(x_1,\ldots,x_m,x)\,dx$$

and successively on the basis of the middle value theorem

$$R(u)=-\left[\frac{d}{dx}u(x_1,\ldots,x_m,x)\right]_{x=\bar{x}}\int_{-1}^{1}\Omega_m(x)\,dx\,,\qquad(\text{with }\bar{x}\in[-1,1])\,.$$

Whence, by virtue of theor. 3.8.II, we get (4.12.26). Formulae (4.12.27) can be obtained using (4.12.24), q.e.d.

Then reasoning similarly as in the end of the preceding paragraph we conclude that *in any case the influence-function $\Phi(x)$ is non negative in* $[-1, 1]$. Successiviley we may deduce the bounding formula of $R(u)$ when $u^{(m+2)}(x)$ (if m is even) or $u^{(m+1)}(x)$ (if m is odd) are bounded in $[a, b]$; namely after having established

$$\int_{-1}^{1} \Phi(x)\, dx = \frac{2}{(m+2)!} \int_{-1}^{1} \Omega_m(x)\, dx = \frac{2\, I_m}{(m+2)!} \qquad (m = 2, 4, 6)\,,$$

$$\int_{-1}^{1} \Phi(x)\, dx = -\frac{1}{(m+1)!} \int_{-1}^{1} \Omega_m(x)\, dx = \frac{I_m}{(m+1)!} \quad (m = 1, 3, 5, 7, 9)\,,$$

we find

$$|R(u)| \leqq \frac{2\, I_m}{(m+2)!} \sup |u^{(2m+2)}(x)|\,, \qquad (m = 2, 4, 6)\,,$$

$$|R(u)| \leqq \frac{I_m}{(m+1)!} \sup |u^{(m+1)}(x)|\,, \qquad (m = 1, 3, 5, 7, 9)\,.$$

4.13. Quadrature formulae connected to s-orthogonal polynomials.[1]) Let us construct a quadrature formula of the type

$$\int_a^b p(x)\, u(x)\, dx = \sum_{h=0}^{r} \sum_{i=1}^{m} A_{hi}\, u^{(h)}(x_i) + R(u)\,, \qquad (r \geqq 0)\,, \quad (4.13.1)$$

with the weight $p(x)$ *non negative* in the interval $[a, b]$ (finite or infinite) and under the condition that it be exact when the argument function $u(x)$ be a polynomial of degree $\leqq m\,(r+2) - 1$. Namely let us assume that

$$E = \frac{d^n}{dx^n} \quad \text{with} \quad n = m\,(r+2)\,. \qquad (4.13.2)$$

It is obvious that there results $r < n - 1$ and therefore the study of (4.13.1) is equivalent to a Gauss problem. On the basis of theorem 2.5.I we must consider the boundary differential problem

$$\frac{d^{m(r+2)} u}{dx^{m(r+2)}} = 0\,; \quad u^{(h)}(x_i) = 0\,, \quad (h = 0, 1, \ldots, r;\ i = 1, 2, \ldots, m)\,.$$

Such a problem has the following linearly independent non trivial solutions $U_k(x) = x^k \prod_{i=1}^{m} (x - x_i)^{r+1}$, $(k = 0, 1, \ldots, m-1)$ and therefore (4.13.1) can hold if and only if the nodes $x_1, x_2, \ldots, x_m$ are chosen in such a way as to give

$$\int_a^b p(x)\, x^k \left[\prod_{i=1}^{m} (x - x_i)\right]^{r+1} dx = 0\,, \quad (k = 0, 1, \ldots, m-1)\,. \quad (4.13.3)$$

It is evident that this may hold if and only if r is an even number; putting $r = 2\,s$ with $s \geqq 0$, formula (4.13.3) is written

$$\int_a^b p(x)\, x^k \left[\prod_{i=1}^{m} (x - x_i)\right]^{2s+1} dx = 0\,, \quad (k = 0, 1, \ldots, m-1),$$

1) See A. Ossicini [3].

whence, comparing it with (3.9.2), we see that $\prod_{i=1}^{m}(x - x_i)$ must coincide with the polynomial $P_{s,m}(x)$ of degree m of the sequence of polynomials s-orthogonal in $[a, b]$ with respect to the weight function $p(x)$. Then the *nodes* $x_1, x_2, \ldots, x_m$ *must be placed at the* m *zeros of the polynomial* $P_{s,m}(x)$, which (theor. 3.9.II) are real, distinct and internal in the interval $[a, b]$.

Choosing the nodes in this way, formula (4.13.1), which here we rewrite for $r = 2s$

$$\boxed{\int_a^b p(x)\, u(x)\, dx = \sum_{h=0}^{2s} \sum_{i=1}^{m} A_{hi}\, u^{(h)}(x_i) + R(u)}\ , \qquad (s \geqq 0)\ , \quad (4.13.4)$$

is possible in one and only one way since (with the notations of theor. 2.5.I) we have $m(n - p) - n + q = m(2s + 1) - m(2s + 2) + m = 0$.

For $s = 0$, we have the already seen quadrature formulae with the nodes in the zeros of the orthogonal polynomials with respect to the weight $p(x)$, of which we have given the most remarkable examples in § 4,5, 4.9, 4.10.

About the hypotheses to do on $p(x)$, $u(x)$ for the validity of (4.13.4), we prove the following theorem:

Theorem 4.13.I. *Formula (4.13.4) is valid under the following hypotheses:*

$$p(x) \in L[a, b]\ , \quad u(x) \in AC^{n-1}[a, b]\ , \quad (\text{if } a, b \text{ are finite})\ ; \qquad (4.13.5)$$

$$\left.\begin{aligned} &x^n\, p(x) \in L[0, +\infty)\ , \quad u(x) \in A\, C^{n-1}_{\text{loc}}[0, +\infty)\ , \\ &u^{(n)}(x) \int_x^{+\infty} \xi^{n-1}\, p(\xi)\, d\xi \in L[0, +\infty)\ , \quad (\text{if } a = 0,\ b = +\infty)\ ;^{1)} \end{aligned}\right\} \qquad (4.13.6)$$

$$\left.\begin{aligned} &x^n\, p(x) \in L(-\infty, +\infty)\ , \quad u(x) \in A\, C^{n-1}_{\text{loc}}(-\infty, +\infty)\ , \\ &u^{(n)}(x) \int_{-\infty}^{x} \xi^{n-1}\, p(\xi)\, d\xi \in L(-\infty, 0]\ , \\ &u^{(n)}(x) \int_x^{+\infty} \xi^{n-1}\, p(\xi)\, d\xi \in L[0, +\infty)\ , \quad (\text{if } a = -\infty,\ b = +\infty)\ . \end{aligned}\right\} \qquad (4.13.7)$$

Proof. Formulae (4.13.5) coincide with (2.1.2).

To obtain (4.13.6) we first must recall that for (3.9.7) we have to do the hypothesis $p(x)\, x^{m(2s+2)} \in L[0, +\infty)$ that is

$$x^n\, p(x) \in L[0, +\infty)\ . \qquad (4.13.8)$$

We must then impose the validity of (2.2.2), (2.2.11), (2.2.12) considering that we have $u_i(x) = \dfrac{x^{i-1}}{(i-1)!}$, $v_i(x) = (-1)^{n-i}\dfrac{x^{n-i}}{(n-i)!}$ (see § 1.5, Problem 1). For $p(x)$ there result the conditions $p(x) \in L_{\text{loc}}[0, +\infty)$; $x^{i-1}\, p(x) \in L[0, +\infty)$, $(i = 1, 2, \ldots, n)$; but these are obvious consequences of (4.13.8). On the contrary for $u(x)$ we find $u(x) \in A\, C^{n-1}_{\text{loc}}[0, +\infty)$; $u^{(n)}(x)\, x^{n-i} \int_x^{+\infty} \xi^{i-1}\, p(\xi)\, d\xi$

[1]) Without loss of generality, we assume $a = 0$.

$\in L[0, +\infty)$, $(i = 1, 2, \ldots, n)$; but for $x \geqq 0$ we have

$$\int_x^{+\infty} \xi^{n-1}\, p(\xi)\, d\xi = \int_x^{+\infty} \xi^{n-i}\, \xi^{i-1}\, p(\xi)\, d\xi \geqq x^{n-i} \int_x^{+\infty} \xi^{i-1}\, p(\xi)\, d\xi$$

and therefore

$$\left| u^{(n)}(x)\, x^{n-i} \int_x^{+\infty} \xi^{i-1}\, p(\xi)\, d\xi \right| \leqq \left| u^{(n)}(x) \int_x^{+\infty} \xi^{n-1}\, p(\xi)\, d\xi \right| ,$$

to that of the preceding n summability conditions it suffices to write only those with $i = n$. We thus obtain (4.13.6). Likewise we obtain (4.13.7) using (3.9.7), (2.3.2), (2.3.8), q.e.d.

For computing the $m\,(2\,s + 1)$ coefficients A_{hi} of the formula (4.13.4) the best way, from the practical point of view, is the following. Writing that (4.13.4) is exact for $u(x) = x^k$, $(k = 0, 1, \ldots, n - 1)$, we get the following system of $n = m\,(2\,s + 2)$ linear equations in the $m\,(2\,s + 1) = n - m$ unknowns A_{hi}:

$$\int_a^b x^k\, p(x)\, dx = \sum_{h=0}^{\min(2s,\,k)} \sum_{i=1}^{m} A_{hi} \frac{k!}{(k-h)!}\, x_i^{k-h}, \qquad (k = 0, 1, \ldots, n-1)\,; \quad (4.13.9)$$

we know that this system has certainly one and only one solution. If in the formulae (4.13.9) we omit the last m equations, we get a system of $n - m$ equations in $n - m$ unknowns and we may prove (see A. Ossicini [3]) that the corresponding determinant of the coefficients is different from zero; solving such a system we obtain the required coefficients A_{hi}.

For instance, in the particular cases considered in § 3.9 we get the following formulae (see L. Rebolia [1], I. Verna [1]):

$$\int_{-1}^{1} u(x)\, dx = u(-\alpha) + u(\alpha) + 0{,}09629177\, [u'(-\alpha) - u'(\alpha)]$$
$$+ 0{,}02930120\, [u''(-\alpha) + u''(\alpha)] + R(u)\,, \qquad (4.13.10)$$

$\alpha = 0{,}6292111$; $R(u) = 0$ if $u(x)$ is a polynomial of degree $\leqq 7$;

$$\int_{-1}^{1} u(x)\, dx = 0{,}5333164\, [u(-\alpha) + u(\alpha)] + 0{,}9333672\, u(0)$$
$$+ 0{,}03116467\, [u'(-\alpha) - u'(\alpha)] + 0{,}004107481\, [u''(-\alpha)$$
$$+ u''(\alpha)] + 0{,}02212708\, u''(0) + R(u)\,, \qquad (4.13.11)$$

$\alpha = 0{,}8144392$; $R(u) = 0$ if $u(x)$ is a polynomial of degree $\leqq 11$;

$$\int_{-1}^{1} u(x)\, dx = 0{,}3246869\, [u(-\alpha) + u(\alpha)] + 0{,}6753131\, [u(-\beta) + u(\beta)]$$
$$+ 0{,}01201753\, [u'(-\alpha) - u'(\alpha)] + 0{,}01019067\, [u'(-\beta) - u'(\beta)]$$
$$+ 0{,}0008997426\, [u''(-\alpha) + u''(\alpha)]$$
$$+ 0{,}008133556\, [u''(-\beta) + u''(\beta)] + R(u)\,, \qquad (4.13.12)$$

$\alpha = 0{,}8896768$; $\beta = 0{,}3588585$;

$R(u) = 0$ if $u(x)$ is a polynomial of degree $\leqq 15$;

$$\int_{-1}^{1} u(x)\,dx = 0{,}2169538\,[u(-\alpha) + u(\alpha)] + 0{,}4878607\,[u(-\beta) + u(\beta)]$$
$$+ 0{,}5903711\,u(0) + 0{,}005461143\,[u'(-\alpha) - u'(\alpha)]$$
$$+ 0{,}007506891\,[u'(-\beta) - u'(\beta)]$$
$$+ 0{,}0002646102\,[u''(-\alpha) + u''(\alpha)]$$
$$+ 0{,}003022541\,[u''(-\beta) + u''(\beta)]$$
$$+ 0{,}005356505\,u''(0) + R(u) \qquad (4.13.13)$$

$\alpha = 0{,}9271179$; $\beta = 0{,}5608674$;

$R(u) = 0$ if $u(x)$ is a polynomial of degree $\leqq 19$;

$$\int_{-1}^{1} u(x)\,dx = u(-\alpha) + u(\alpha) + 0{,}1288705\,[u'(-\alpha) - u'(\alpha)]$$
$$+ 0{,}03916800\,[u''(-\alpha) + u''(\alpha)]$$
$$+ 0{,}002619962\,[u'''(-\alpha) - u'''(\alpha)]$$
$$+ 0{,}0002216673\,[u''''(-\alpha) + u''''(\alpha)] + R(u)\,, \qquad (4.13.14)$$

$\alpha = 0{,}6500278$; $R(u) = 0$ if $u(x)$ is a polynomial of degree $\leqq 11$;

$$\int_{-1}^{1} u(x)\,dx = 0{,}6099488\,[u(-\alpha) + u(\alpha)] + 0{,}7801025\,u(0)$$
$$+ 0{,}05236188\,[u'(-\alpha) - u'(\alpha)] + 0{,}008865985\,[u''(-\alpha) + u''(\alpha)]$$
$$+ 0{,}01242089\,u''(0) + 0{,}0003732575\,[u'''(-\alpha) - u'''(\alpha)]$$
$$+ 0{,}00001679051\,[u''''(-\alpha) + u''''(\alpha)] + R(u)\,,\ ^{1)} \qquad (4.13.15)$$

$\alpha = 0{,}7960777$; $R(u) = 0$ if $u(x)$ is a polynomial of degree $\leqq 17$;

$$\int_{-1}^{1} u(x)\,dx = 0{,}3156042\,[u(-\alpha) + u(\alpha)] + 0{,}6843958\,[u(-\beta) + u(\beta)]$$
$$+ 0{,}01517919\,[u'(-\alpha) - u'(\alpha)] + 0{,}01355609\,[u'(-\beta) - u'(\beta)]$$
$$+ 0{,}001213977\,[u''(-\alpha) + u''(\alpha)] + 0{,}01048016\,[u''(-\beta) + u''(\beta)]$$
$$+ 0{,}00002674037\,[u'''(-\alpha) - u'''(\alpha)]$$
$$+ 0{,}0001128025\,[u'''(-\beta) - u'''(\beta)]$$
$$+ 0{,}0000005426435\,[u''''(-\alpha) + u''''(\alpha)]$$
$$+ 0{,}00002636424\,[u''''(-\beta) + u''''(\beta)] + R(u)\,, \qquad (4.13.16)$$

$\alpha = 0{,}8998292$; $\beta = 0{,}3659244$; $R(u) = 0$ if $u(x)$ is a polynomial of degree $\leqq 23$.

[1] We have omitted the therm in $u''''(0)$ since its coefficient results $< 10^{-20}$.

$$\begin{aligned}\int_{-1}^{1} u(x)\,dx &= 0{,}2093506\,[u(-\alpha)+u(\alpha)] + 0{,}4913607\,[u(-\beta)+u(\beta)]\\ &+ 0{,}5985774\,u(0) + 0{,}006796213\,[u'(-\alpha)-u'(\alpha)]\\ &+ 0{,}009834182\,[u'(-\beta)-u'(\beta)]\\ &+ 0{,}0003535743\,[u''(-\alpha)+u''(\alpha)]\\ &+ 0{,}003870873\,[u''(-\beta)+u''(\beta)] + 0{,}006902554\,u''(0)\\ &+ 0{,}000005175554\,[u'''(-\alpha)-u'''(\alpha)]\\ &+ 0{,}00004143548\,[u'''(-\beta)-u'''(\beta)]\\ &+ 0{,}00000006753689\,[u''''(-\alpha)+u''''(\alpha)]\\ &+ 0{,}000004871802\,[u''''(-\beta)+u''''(\beta)]\\ &+ 0{,}00001307175\,u''''(0) + R(u)\,, \end{aligned} \tag{4.13.17}$$

$$\alpha = 0{,}9343690\,; \qquad \beta = 0{,}5690324\,;$$

$$R(u) = 0 \text{ if } u(x) \text{ is a polynomial of degree } \leqq 29\,;$$

$$\int_{0}^{+\infty} e^{-x}\,u(x)\,dx = u(\alpha) - 0{,}5960716\,u'(\alpha) + 0{,}6776507\,u''(\alpha) + R(u)\,; \tag{4.13.18}$$

$$\alpha = 1{,}5960716\,; \qquad R(u) = 0 \text{ if } u(x) \text{ is a polynomial of degree } \leqq 3\,;$$

$$\begin{aligned}\int_{0}^{+\infty} e^{-x}\,u(x)\,dx &= 0{,}9646119\,u(\alpha) + 0{,}03538813\,u(\beta) - 0{,}03768718\,u'(\alpha)\\ &- 0{,}05024515\,u'(\beta) + 0{,}2172137\,u''(\alpha)\\ &+ 0{,}03038297\,u''(\beta) + R(u)\,,\end{aligned} \tag{4.13.19}$$

$$\alpha = 0{,}8934994\,; \qquad \beta = 6{,}3877983\,;$$

$$R(u) = 0 \text{ if } u(x) \text{ is a polynomial of degree } \leqq 7\,;$$

$$\begin{aligned}\int_{0}^{+\infty} e^{-x}\,u(x)\,dx &= 0{,}8824642\,u(\alpha) + 0{,}1173281\,u(\beta) + 0{,}0002076883\,u(\gamma)\\ &+ 0{,}05308019\,u'(\alpha) - 0{,}09555721\,u'(\beta) - 0{,}0003378884\,u'(\gamma)\\ &+ 0{,}09356091\,u''(\alpha) + 0{,}06068781\,u''(\beta)\\ &+ 0{,}0001895134\,u''(\gamma) + R(u)\,,\end{aligned} \tag{4.13.20}$$

$$\alpha = 0{,}6212774\,; \qquad \beta = 4{,}1936031\,; \qquad \gamma = 12{,}1938097\,;$$

$$R(u) = 0 \text{ if } u(x) \text{ is a polynomial of degree } \leqq 11;$$

$$\begin{aligned}\int_{-\infty}^{+\infty} e^{-x^2}\,u(x)\,dx &= 0{,}8862269\,[u(-\alpha)+u(\alpha)]\\ &+ 0{,}3215879\,[u'(-\alpha)-u'(\alpha)]\\ &+ 0{,}08479382\,[u''(-\alpha)+u''(\alpha)] + R(u)\,,\end{aligned} \tag{4.13.21}$$

$$\alpha = 1{,}0264374\,; \qquad R(u) = 0 \text{ if } u(x) \text{ is a polynomial of degree } \leqq 7\,;$$

$$\int_{-\infty}^{+\infty} e^{-x^2} u(x)\, dx = 0{,}1551204\, [u(-\alpha) + u(\alpha)] + 1{,}4622131\, u(0) + 0{,}05195775\, [u'(-\alpha) - u'(\alpha)] + 0{,}007837287\, [u''(-\alpha) + u''(\alpha)] + 0{,}1253966\, u''(0) + R(u)\,, \qquad (4.13.22),$$

$\alpha = 1{,}7699914$; $R(u) = 0$ if $u(x)$ is a polynomial of degree $\leqq 11$;

$$\int_{-\infty}^{+\infty} e^{-x^2} u(x)\, dx = 0{,}8862269\, [u(-\alpha) + u(\alpha)] + 0{,}5451873\, [u'(-\alpha) - u'(\alpha)] + 0{,}2000577\, [u''(-\alpha) + u''(\alpha)] + 0{,}03739656\, [u'''(-\alpha) - u'''(\alpha)] + 0{,}004026832\, [u''''(-\alpha) + u''''(\alpha)] + R(u)\,, \qquad (4.13.23)$$

$\alpha = 1{,}2686013$; $R(u) = 0$ if $u(x)$ is a polynomial of degree $\leqq 11$;

$$\int_{-\infty}^{+\infty} e^{-x^2} u(x)\, dx = 0{,}08600773\, [u(-\alpha) + u(\alpha)] + 1{,}6004384\, u(0) + 0{,}04817540\, [u'(-\alpha) - u'(\alpha)] + 0{,}01259816\, [u''(-\alpha) + u''(\alpha)] + 0{,}2180515\, u''(0) + 0{,}001668321\, [u'''(-\alpha) - u'''(\alpha)] + 0{,}0001027652\, [u''''(-\alpha) + u''''(\alpha)] + 0{,}006576976\, u''''(0) + R(u)\,, \qquad (4.13.24)$$

$\alpha = 2{,}1841838$; $R(u) = 0$ if $u(x)$ is a polynomial of degree $\leqq 17$.

Considering again the general case and assuming already computed the coefficients A_{hi}, we may deduce the expression of the remainder $R(u)$ applying as usually theorem 2.4.I. Namely we obtain:

$$R(u) = \int_a^b \Phi(x)\, u^{(n)}(x)\, dx\,, \qquad (4.13.25)$$

where the influence-function $\Phi(x)$ is expressed by

$$\Phi(x) = \varphi_i(x) \quad \text{for} \quad x_i < x < x_{i+1}\,, \quad (i = 0, 1, \ldots, m;\, x_0 = a,\, x_{m+1} = b)\,, \qquad (4.13.26)$$

and the functions $\varphi_i(x)$, integrals of the differential equation $\varphi^{(n)}(x) = p(x)$ (since n is even), are given by the formulae

$$\varphi_i(x) = \int_a^x p(\xi) \frac{(x-\xi)^{n-1}}{(n-1)!}\, d\xi - \sum_{h=0}^{2s} \sum_{j=1}^{i} (-1)^h A_{hj} \frac{(x-x_j)^{n-h-1}}{(n-h-1)!}\,, \text{[1)]} \qquad (4.13.27)$$

1) For $i = m$ we may also write $\varphi_m(x) = -\int_x^b p(\xi) \frac{(x-\xi)^{n-1}}{(n-1)!}\, d\xi$.

We now prove the following theorems:

Theorem 4.13.II. *Under the hypothesis that the weight $p(x)$ be not identically zero in any interval contained in $[a, b]$, the influence function $\Phi(x)$ defined by (4.13.26) [together with (4.13.27)] belongs to the class C^{n-2s-2} in $[a, b]$ and is positive inside such interval.*[1])

Proof. From (4.13.26), (4.13.27) it follows, differentiating k times (with $0 \leqq k \leqq n-1$):

$$\Phi^{(k)}(x) = \varphi_i^{(k)}(x) \quad \text{for } x_i < x < x_{i+1}, \quad (i = 0, 1, \ldots, m), \tag{4.13.28}$$

$$\varphi_i^{(k)}(x) = \int_a^x p(\xi) \frac{(x-\xi)^{n-k-1}}{(n-k-1)!} d\xi - \sum_{h=0}^{2s} \sum_{j=1}^{i} (-1)^h A_{hj} \frac{(x-x_j)^{n-h-k-1}}{(n-h-k-1)!},$$
$$(0 \leqq k \leqq n-2s-2), \tag{4.13.29}$$

$$\varphi_i^{(k)}(x) = \int_a^x p(\xi) \frac{(x-\xi)^{n-k-1}}{(n-k-1)!} d\xi - \sum_{h=0}^{n-k-1} \sum_{j=1}^{i} (-1)^h A_{hj} \frac{(x-x_j)^{n-h-k-1}}{(n-h-k-1)!},$$
$$(n-2s-1 \leqq k \leqq n-1).\,[2]) \tag{4.13.30}$$

From the above formulae, keeping into account the hypotheses done on the weight $p(x)$, we see that:

$$\Phi^{(k)}(a) = \Phi^{(k)}(b) = 0, \tag{4.13.31}$$

$$\Phi^{(k)}(x) > 0 \quad \text{for } a < x < x_1, \quad (-1)^k \Phi^{(k)}(x) > 0 \quad \text{for } x_m < x < b, \tag{4.13.32}$$
$$(k = 0, 1, \ldots, n-1).$$

Moreover, since in (4.13.29) we have $n - h - k - 1 \geqq n - 2s - (n - 2s - 2) - 1 = 1$, it is clear that the functions $\Phi(x)$, $\Phi'(x)$, ..., $\Phi^{(n-2s-2)}(x)$ are continuous in $[a, b]$, while, since in (4.13.30) we have $n - h - k - 1 \geqq 0$, $\Phi^{(n-2s-1)}(x)$, ..., $\Phi^{(n-1)}(x)$ have discontinuities of first kind at the points $x_1, x_2, \ldots, x_m$.

We show that $\Phi^{(n-2s-2)}(x)$ *has at most* $2s+2$ *zeros in each interval* $[x_i, x_{i+1}]$, $(i = 1, 2, \ldots, m-1)$. In fact, should it have $2s+3$ of them, for the Rolle theorem, $\Phi^{(n-2s-1)}(x)$ would have at least $2s+2$ zeros inside $[x_i, x_{i+1}]$, $\Phi^{(n-2s)}(x)$ would have at least $2s+1$ zeros and so on, until we may conclude that $\Phi^{(n-1)}(x)$ would have at least two zeros inside $[x_i, x_{i+1}]$. But this is absurd since from (4.13.28), (4.13.30) there follows that, for $x_i < x < x_{i+1}$, we have $\Phi^{(n-1)}(x) = \varphi_i^{(n-1)}(x) = \int_a^x p(\xi)\,d\xi - \sum_{j=1}^{i} A_{0j}$ and this

[1]) For $s = 0$ this theorem includes as particular cases the theorem 4.5.I, the theorem 4.9.I and the formulae (4.10.10).

[2]) For $i = m$ the two last formulae can be jointed in the only one

$$\varphi_m^{(k)}(x) = -\int_x^b p(\xi) \frac{(x-\xi)^{n-k-1}}{(n-k-1)!} d\xi.$$

function is *increasing* [for the hypothesis on $p(x)$ given in the statement]. On the basis of the result now obtained and recalling (4.13.32) we may obviously deduce that *$\Phi^{(n-2s-2)}(x)$ has at most* $(m-1)(2s+2) = n-2s-2$ *zeros inside* $[a, b]$.

We may then show that *$\Phi(x)$ does not vanish inside* $[a, b]$ *and therefore* is positive, by virtue of (4.13.32). In fact, if $\Phi(x)$ should vanish at one point in (a, b), using (4.13.31) and applying Rolle theorem, we find that $\Phi^{(n-2s-2)}(x)$ would vanish at least $n-2s-1$ times, in contraposition with the preceding deduction. The theorem is thus proved.

Theorem 4.13.III. *Assuming that the polynomial $P_{s,m}(x)$ be normalized in such a way that*

$$P_{s,m}(x) = x^m + \cdots \quad \text{or} \quad P_{s,m}(x) = (x-x_1)(x-x_2)\ldots(x-x_m)\,, \tag{4.13.33}$$

then there results

$$\int_a^b \Phi(x)\,dx = \frac{1}{n!}\int_a^b p(x)\,[P_{s,m}(x)]^{2s+2}\,dx\,, \qquad [n = m(2s+2)]\,. \tag{4.13.34}$$

The same integral can also be expressed in the following way

$$\int_a^b \Phi(x)\,dx = \frac{1}{n!}\int_a^b x^n\,p(x)\,dx - \sum_{h=0}^{2s}\sum_{i=1}^{m} A_{hi}\frac{x_i^{n-h}}{(n-h)!}\,. \tag{4.13.35}$$

Proof. If we put in (4.13.4) $u(x) = [P_{s,m}(x)]^{2s+2} = (x^m + \cdots)^{2s+2} = x^n + \cdots$ and we keep into account the expression (4.13.25) of $R(u)$, we get

$$\int_a^b p(x)\,[P_{s,m}(x)]^{2s+2}\,dx = \int_a^b \Phi(x)\,n!\,dx$$

from which we have (4.13.34). Formula (4.13.35) can be soon obtained by putting $u(x) = x^n$.

Theorem 4.13.IV. *In the particular case $a = -1$, $b = 1$, $p(x) = 1$, we have*

$$\int_{-1}^{1} \Phi(x)\,dx = \frac{2}{(n+1)!}\,[(1-x_1)(1-x_2)\ldots(1-x_m)]^{2s+2}\,. \tag{4.13.36}$$

Proof. If (4.13.33) holds, the polynomial $\overline{P}_{s,m}(x) = \dfrac{P_{s,m}(x)}{(1-x_1)(1-x_2)\ldots(1-x_m)}$ is normalized in such a way that $\overline{P}_{s,m}(1) = 1$.

We then know from (3.9.11) that there results

$$\int_{-1}^{1} [\overline{P}_{s,m}(x)]^{2s+2}\,dx = \frac{2}{1+(2s+2)m} = \frac{2}{n+1}\,.$$

There follows

$$\int_{-1}^{1} [P_{s,m}(x)]^{2s+2}\, dx = [(1-x_1)\ldots(1-x_m)]^{2s+2}\cdot\frac{2}{n+1}$$

and substituting in (4.13.34) we obtain (4.13.36).

From (4.13.25), using theorems 4.13.II and 4.13.III we may get the usual bounding formulae of the remainder $R(u)$. We leave the easy task to the reader, who will use also (4.13.36) to bound the remainder of the formulae (4.13.10), (4.13.11), ..., (4.13.17).

4.14. Some examples of quadrature formulae exact for trigonometric polynomials.[1]) We want to construct a quadrature formula of the type

$$\int_{-\pi}^{\pi} u(x)\, dx = \sum_{h=0}^{2s}\sum_{j=1}^{m} A_{hj}\, u^{(h)}(x_j) + R(u)\,, \quad (s \geqq 0)\,, \tag{4.14.1}$$

which be exact when $u(x)$ is a trigonometric polynomial of order $m(s+1)-1$. Namely we assume that it be

$$E = \frac{d}{dx}\prod_{k=1}^{m(s+1)-1}\left(\frac{d^2}{dx^2}+k^2\right) \tag{4.14.2}$$

and since this operator has order

$$n = m(2s+2) - 1 \tag{4.14.3}$$

we must do the hypothesis $u(x) \in A\, C^{n-1}[-\pi, \pi]$.

The Gauss problem corresponding to (4.14.1), on the basis of theorem 2.5.I, needs that the following boundary differential problem be taken into consideration:

$$\left.\begin{aligned} &\frac{d}{dx}\prod_{k=1}^{m(s+1)-1}\left(\frac{d^2}{dx^2}+k^2\right)u = 0\,; \qquad u^{(h)}(x_j) = 0\,; \\ &(h = 0, 1, \ldots, 2s;\, j = 1, 2, \ldots, m)\,. \end{aligned}\right\} \tag{4.14.4}$$

We shall soon see that this problem has only the trivial solution if $m = 1$, while he has non-trivial solutions if $m > 1$. Therefore it is convenient to distinguish since now the two cases $m = 1$ and $m > 1$. If $m = 1$ we shall exclude the common case $s = 0$.

In the case $m = 1$, $s > 0$ the solution of (4.14.4) must be a trigonometric polynomial of order s with a zero of multiplicity $2s + 1$ at the point x_1. Such a polynomial is necessarily identically zero;[2]) therefore the problem (4.14.4) has no non-trivial solutions and then *if $m = 1$ the quadrature formula (4.14.1) is possible whatever is $x_1 \in [-\pi, \pi]$*. Having fixed x_1, it is unique

[1]) See F. Rosati [1].

[2]) It suffices to consider that, if we put $= 0$ a trigonometric polynomial in x, of degree s, we obtain an equation equivalent to an algebraic equation in $z = e^{ix}$, of degree $2s$.

since, using the notations of theorem 2.5.I, we have $m(n-p)-n=1.(2s+1)-(2s+1)=0$.

To compute the coefficients A_{h1}, it is convenient to remark that the expression $\sum_{h=0}^{2s} A_{h1} u^{(h)}(x_1)$ must vanish for $u=\cos kx, \sin kx, (k=1,2,\ldots,s)$ whatewer is x_1, since in (4.14.1) it must be $R(u)=0$ when $u(x)$ coincides with one of the functions $\cos kx$, $\sin kx$, $(k=1,2,\ldots,s)$, for which we have $\int_{-\pi}^{\pi} u(x)\,dx=0$. Evidently this is possible only if

$$\sum_{h=0}^{2s} A_{h1} u^{(h)}(x_1) = c\left[\prod_{k=1}^{s}\left(\frac{d^2}{dx^2}+k^2\right)u(x)\right]_{x=x_1}$$

with c constant. Then writing that (4.14.1) (with $m=1$) is exact for $u(x)=1$ we obtain $2\pi = c\prod_{k=1}^{s} k^2$ from which $c=\frac{2\pi}{(s!)^2}$. Therefore *in the case* $m=1$, *(4.14.1) can be written*

$$\boxed{\int_{-\pi}^{\pi} u(x)\,dx = \frac{2\pi}{(s!)^2}\left[\prod_{k=1}^{s}\left(\frac{d^2}{dx^2}+k^2\right)u(x)\right]_{x=x_1} + R(u)} \qquad (4.14.5)$$

where x_1 *is an arbitrary point of* $[-\pi,\pi]$.

For to obtain the remainder $R(u)$, it suffices to apply theorem 2.4.I thus we get

$$R(u) = \int_{-\pi}^{\pi} \Phi(x)\cdot\frac{d}{dx}\prod_{k=1}^{s}\left(\frac{d^2}{dx^2}+k^2\right)u(x)\,dx\,, \qquad (4.14.6)$$

where the influence-function $\Phi(x)$ is given by

$$\Phi(x) = \begin{cases} \varphi_0(x) = -\dfrac{2^{2s}}{(2s)!}\displaystyle\int_{-\pi}^{x}\sin^{2s}\frac{x-\xi}{2}\,d\xi\,, & \text{for } -\pi<x<x_1\,, \\[2ex] \varphi_1(x) = \dfrac{2^{2s}}{(2s)!}\displaystyle\int_{x}^{\pi}\sin^{2s}\frac{x-\xi}{2}\,d\xi\,, & \text{for } x_1<x<\pi\,, \end{cases} \qquad (4.14.7)$$

having considered that the resolvent kernel $K(x,\xi)$ relative to the differential operator (4.14.2) with $m=1$ is given by

$$K(x,\xi) = \frac{2^{2s}}{(2s)!}\sin^{2s}\frac{x-\xi}{2}\,.\text{ [1]}$$

[1] Formula (4.14.7) is written under the hypothesis that $-\pi<x_1<\pi$; it is obvious how it must be modified if $x_1=\pm\pi$.

In the case $m > 1$, $s \geqq 0$ it is easy to see that, introducing the following trigonometric polynomials $\left[\text{of order } (2s+1)\frac{m}{2} + \frac{m}{2} - 1 - h = m(s+1) - 1 - h\right]$:

$$U_h(x) = \left(\prod_{j=1}^{m} \sin\frac{x - x_j}{2}\right)^{2s+1} \cos\left(\frac{m}{2} - 1 - h\right)x\,, \qquad \left(0 \leqq h \leqq \frac{m}{2} - 1\right),$$
$$V_h(x) = \left(\prod_{j=1}^{m} \sin\frac{x - x_j}{2}\right)^{2s+1} \sin\left(\frac{m}{2} - 1 - h\right)x\,,$$

the boundary problem (4.14.4) has the following $m - 1$ linearly independent non trivial solutions

$$U_h(x),\ V_h(x) \quad \text{with} \quad h = 0, 1, \ldots, \frac{m-1}{2} - 1\,, \qquad (\text{if } m \text{ is odd})\,,$$

$$U_h(x) \quad \text{with } h = 0, 1, \ldots, \frac{m}{2} - 1\,; \quad V_h(x) \quad \text{with} \quad h = 0, 1, \ldots, \frac{m}{2} - 2\,,$$
$$(\text{if } m \text{ is even})\,.$$

Therefore, for theorem 2.5.I, the quadrature formula (4.14.1) may hold if and only if the nodes $x_1, x_2, \ldots, x_m$ are chosen in such a way as to give

$$\int_{-\pi}^{\pi} \left(\prod_{j=1}^{m} \sin\frac{x - x_j}{2}\right)^{2s+1} \cos\left(\frac{m}{2} - 1 - h\right)x\, dx = 0\,,$$

$$\left(h = 0, 1, \ldots, \begin{matrix} \frac{m-1}{2} - 1 \\ \frac{m}{2} - 1 \end{matrix}\right),$$

$$\int_{-\pi}^{\pi} \left(\prod_{j=1}^{m} \sin\frac{x - x_j}{2}\right)^{2s+1} \sin\left(\frac{m}{2} - 1 - h\right)x\, dx = 0\,,$$

$$\left(h = 0, 1, \ldots, \begin{matrix} \frac{m-1}{2} - 1 \\ \frac{m}{2} - 2 \end{matrix}\right),$$

or more concisely

$$\int_{-\pi}^{\pi} \left(\prod_{j=1}^{m} \sin\frac{x - x_j}{2}\right)^{2s+1} \Pi_{\frac{m}{2}-1}(x)\, dx = 0 \qquad (4.14.8)$$

where $\Pi_{\frac{m}{2}-1}(x)$ denotes an arbitrary trigonometric polynomial of order $\frac{m}{2} - 1$.[1])

It is easy to verify that (4.14.8) is satisfied when the trigonometric polynomial $\prod_{j=1}^{m} \sin \frac{x - x_j}{2}$ $\left(\text{of order } \frac{m}{2}\right)$ is of the type $a \cos \frac{m}{2} x + b \sin \frac{m}{2} x$. In fact, using for instance the representation with imaginary exponentials, we see that $\left(a \cos \frac{m}{2} x + b \sin \frac{m}{2} x\right)^{2s+1}$ is a linear combination of the functions

$\cos \frac{m}{2} x$, $\cos 3 \frac{m}{2} x, \ldots, \cos (2s+1) \frac{m}{2} x$, $\sin \frac{m}{2} x$, $\sin 3 \frac{m}{2} x, \ldots, \sin (2s+1) \frac{m}{2} x$ each one of which is evidently orthogonal to $\Pi_{\frac{m}{2}-1}(x)$ in the interval $[-\pi, \pi]$.

But we can also prove the contrary and namely that *if* $\prod_{j=1}^{m} \sin \frac{x - x_j}{2}$ *satisfies (4.14.8), necessarily we have*

$$\prod_{j=1}^{m} \sin \frac{x - x_j}{2} = a \cos \frac{m}{2} x + b \sin \frac{m}{2} x\,, \tag{4.14.9}$$

with a, b arbitrary constants.

In fact, we may in every case put

$$\prod_{j=1}^{m} \sin \frac{x - x_j}{2} = a \cos \frac{m}{2} x + b \sin \frac{m}{2} x + P_{\frac{m}{2}-1}(x) \tag{4.14.10}$$

where $P_{\frac{m}{2}-1}(x)$ denotes a trigonometric polynomial of order $\frac{m}{2} - 1$. Substituting (4.14.10) in (4.14.8) we obtain

$$\int_{-\pi}^{\pi} \left[a \cos \frac{m}{2} x + b \sin \frac{m}{2} x + P_{\frac{m}{2}-1}(x)\right]^{2s+1} \Pi_{\frac{m}{2}-1}(x)\, dx = 0\,,$$

which in particular must be fulfilled for $\Pi_{\frac{m}{2}-1}(x) = P_{\frac{m}{2}-1}(x)$; therefore there must hold

$$\int_{-\pi}^{\pi} \left[a \cos \frac{m}{2} x + b \sin \frac{m}{2} x + P_{\frac{m}{2}-1}(x)\right]^{2s+1} P_{\frac{m}{2}-1}(x)\, dx = 0\,. \tag{4.14.11}$$

But, for the previous proof, we also have

$$\int_{-\pi}^{\pi} \left(a \cos \frac{m}{2} x + b \sin \frac{m}{2} x\right)^{2s+1} P_{\frac{m}{2}-1}(x)\, dx = 0 \tag{4.14.12}$$

[1]) From now on, we shall consider not only trigonometric polynomials of *integer* order, but also of *semi-integer* order $\frac{2k+1}{2}$ $\left(\text{linear combinations of cosines and sines of the arguments } \frac{x}{2}, \frac{3x}{2}, \frac{5x}{2}, \ldots, \frac{(2k+1)x}{2}\right)$.

and therefore, subtracting member by member (4.14.11), (4.14.12) we deduce that it must be

$$\int_{-\pi}^{\pi} \left[P_{\frac{m}{2}-1}(x)\right]^2 \sum_{k=0}^{2s} \left(a \cos \frac{m}{2} x + b \sin \frac{m}{2} x\right)^{2s-k} \left[P_{\frac{m}{2}-1}(x)\right]^k dx = 0\,. \quad (4.14.13)$$

In this integral the integrand function does not change its sign in $[-\pi, \pi]$ and therefore (4.14.13) may hold only if $P_{\frac{m}{2}-1}(x) \equiv 0$; in such a way (4.14.10) reduces to (4.14.9), q.e.d.

Having stated (4.14.9), we see that the nodes $x_1, x_2, \ldots, x_m$ must coincide with the zeros of any trigonometric polynomial of the type $a \cos \frac{m}{2} x + b \sin \frac{m}{2} x$; but such a polynomial has in $[-\pi, \pi]$ precisely m zeros in arithmetic progression of ratio $\frac{2\pi}{m}$ and therefore we may conclude *that (4.14.1) with* $m > 1$ *may hold only if the nodes are given by*

$$x_j = x_1 + (j-1)\frac{2\pi}{m}, \qquad (j = 1, 2, \ldots, m)\,, \qquad (4.14.14)$$

with the first node x_1 *chosen arbitrarily in the interval* $\left[-\pi, -\pi + \frac{2\pi}{m}\right]$.
For any fixed x_1, the formula is then unique since, with the usual notations of theorem 2.5.I, we have

$$m(n-p) - n + q = m(2s+1) - [m(2s+2) - 1] + (m-1) = 0\,.$$

Let us now compute the coefficients A_{hj} of (4.14.1). This formula must be exact for $u(x) = 1$, $\cos rx$, $\sin rx$ with $1 \leq r \leq m(s+1) - 1$ and in particular when

$$u(x) = \cos kmx\,, \quad \sin kmx\,, \qquad (k = 1, 2, \ldots, s)\,. \quad (4.14.15)$$

Limiting ourselves, for the moment, to impose that (4.14.1) be exact for the functions (4.14.15), we find that it must be

$$0 = \sum_{j=1}^{m} \sum_{h=0}^{2s} A_{hj} \begin{bmatrix} (\cos kmx)^{(h)} \\ (\sin kmx)^{(h)} \end{bmatrix}_{x = x_1 + (j-1)\frac{2\pi}{m}}, \qquad (k = 1, 2, \ldots, s)\,,$$

and this is certainly satisfied if we take the coefficients A_{hj} in such a way that there holds the *identity*:

$$\sum_{h=0}^{2s} A_{hj} \frac{d^h}{dx^h} = c \prod_{k=1}^{s} \left(\frac{d^2}{dx^2} + k^2 m^2\right).$$

This suggest us to inquire if it is possible to write (4.14.1) in the following form

$$\int_{-\pi}^{\pi} u(x)\,dx = c \sum_{j=1}^{m} \left[\prod_{k=1}^{s} \left(\frac{d^2}{dx^2} + k^2 m^2\right) u(x)\right]_{x = x_1 + (j-1)\frac{2\pi}{m}} + R(u)\,. \qquad (4.14.16)$$

Assuming that it be exact for $u(x) = 1$ we find $2\pi = c\, m \prod_{k=1}^{s} k^2 m^2$ from which we obtain

$$c = \frac{2\pi}{m} \frac{1}{(m^s s!)^2}. \tag{4.14.17}$$

We have only to verify whether (4.14.16) is or is not exact for the remaining functions $\cos r x$, $\sin r x$ with $0 < r \leqq m(s+1) - 1$ and r *not multiple of* m, namely whether there hold or not

$$0 = c \sum_{j=1}^{m} \left[\prod_{k=1}^{s} (k^2 m^2 - r^2) {}^{\cos r x}_{\sin r x} \right]_{x = x_1 + (j-1)\frac{2\pi}{m}},$$

that is, considering that for the already seen values of r, we certainly have $\prod_{k=1}^{s} (k^2 m^2 - r^2) \neq 0$:

$$\sum_{j=1}^{m} {}^{\cos}_{\sin}\, r \left[x_1 + (j-1) \frac{2\pi}{m} \right] = 0.$$

Now this formula does hold since, for every relative integer r not multiple of m, we have

$$\sum_{j=1}^{m} e^{ir\left[x_1 + (j-1)\frac{2\pi}{m}\right]} = e^{irx_1} \sum_{j=1}^{m} \left(e^{ir\frac{2\pi}{m}} \right)^{j-1} = e^{irx_1} \frac{\left(e^{ir\frac{2\pi}{m}} \right)^m - 1}{e^{ir\frac{2\pi}{m}} - 1} = 0.$$

Therefore there follows from (4.14.16), (4.14.17) that for $m > 1$ the required formula is

$$\boxed{\int_{-\pi}^{\pi} u(x)\, dx = \frac{2\pi}{m} \frac{1}{(m^s s!)^2} \sum_{j=1}^{m} \left[\prod_{k=1}^{s} \left(\frac{d^2}{dx^2} + k^2 m^2 \right) u(x) \right]_{x = x_1 + (j-1)\frac{2\pi}{m}} + R(u)}\,.$$ [1]

(4.14.18)

Comparing this formula with (4.14.1) we see that the coefficients A_{hj} with h odd are zeros and that the coefficients A_{hj} with h even are independent of j. Therefore, writing $2h$ instead of h and $\frac{2\pi}{m} \frac{1}{(m^s s!)^2} c_h$ instead of $A_{2h,j}$ we

[1]) Note that for $m = 1$, (4.14.18) reduces to (4.14.5); note also that for $s = 0$ (4.14.18) becomes

$$\int_{-\pi}^{\pi} u(x)\, dx = \frac{2\pi}{m} \sum_{j=1}^{m} u \left[x_1 + (j-1) \frac{2\pi}{m} \right] + R(u).$$

may write the formula in the following way:

$$\int_{-\pi}^{\pi} u(x)\,dx = \frac{2\pi}{m}\frac{1}{(m^s\,s!)^2}\sum_{h=0}^{s} c_h \sum_{j=1}^{m} u^{(2h)}\left[x_1 + (j-1)\frac{2\pi}{m}\right] + R(u)\,, \quad (4.14.19)$$

with the coefficients c_h defined by the following identities in t:

$$\prod_{k=1}^{s} (t + k^2 m^2) = \sum_{h=0}^{s} c_h t^h\,. \quad (4.14.20)$$

From (4.14.19) we see that *our formula yields also the solution of a Tchebyschef problem* (see § 2.6).

About the remainder $R(u)$, theorem 2.4.I gives

$$R(u) = \int_{-\pi}^{\pi} \Phi(x)\frac{d}{dx}\prod_{k=1}^{m(s+1)-1}\left(\frac{d^2}{dx^2} + k^2\right) u(x)\,dx \quad (4.14.21)$$

with

$$\Phi(x) = \varphi_i(x), \quad x_i < x < x_{i+1}\,, \quad (i = 0, 1, \ldots, m;\, x_0 = -\pi,\, x_{m+1} = \pi)\,.\,^{1)} \quad (4.14.22)$$

For $\varphi_i(x)$ the following expression

$$\varphi_i(x) = -\int_{-\pi}^{x} K(\xi, x)\,d\xi + \frac{2\pi}{m}\frac{1}{(m^s\,s!)^2}\sum_{h=0}^{s} c_h \sum_{j=1}^{i}\left[\frac{\partial^{2h}}{\partial\xi^{2h}} K(\xi, x)\right]_{\xi = x_1 + (j-1)\frac{2\pi}{m}} \quad (4.14.23)$$

holds. In this formula $K(x, \xi)$ denotes the resolvent kernel relative to the operator (4.14.2) and therefore we have

$$K(\xi, x) = K(x, \xi) = -\frac{2^{m(2s+2)-2}}{[m(2s+2)-2]!}\sin^{m(2s+2)-2}\frac{x-\xi}{2}\,. \quad (4.14.24)$$

From the formula (4.14.21) we can get some bounds for $R(u)$, however considering that the influence function $\Phi(x)$ does not have constant sign in $[-\pi, \pi]$. We leave to the reader the task of developping the computations for the first values of m and of s.

4.15. Problems.

1. Obtain the formula of the circumscribed trapezium (4.2.1), assuming $E = \dfrac{d}{dx}$.

2. Obtain the formula of the inscribed trapezium (4.2.6), assuming $E = \dfrac{d}{dx}$.

[1]) It may be $x_0 = x_1$ (if we choose $x_1 = -\pi$) or $x_m = x_{m+1}$ $\left(\text{if we choose } x_1 = -\pi + \dfrac{2\pi}{m}\right)$.

3. Obtain the Cavalieri-Simpson formula (4.2.11) assuming $E = \frac{d}{dx}$; or $E = \frac{d^2}{dx^2}$; or $E = \frac{d^3}{dx^3}$.

4. Prove that for the coefficients $H_{mj}^{(\alpha,\beta)}$ of the Gauss-Jacobi formula, the following relation

$$\frac{1}{H_{mj}^{(\alpha,\beta)}} = \sum_{k=0}^{m-1} \frac{[P_k^{(\alpha,\beta)}(x_{mj}^{(\alpha,\beta)})]^2}{h_k^{(\alpha,\beta)}} \tag{1}$$

holds.

5. Write and explain the Bouzitat formula of second kind, for $\alpha = \beta = \lambda - \frac{1}{2}$.

6. Express the coefficients A_{mj} of the Newton-Cotes formula [see (4.11.8)] by means of the Stirling numbers of first kind S_{im}, $(i = 0, 1, \ldots, m - 1)$, defined by the following identity

$$\prod_{i=1}^{m} (t - i + 1) = \sum_{i=0}^{m-1} S_{im}\, t^{m-i}, \tag{1}$$

or from the following recursive process

$$\left.\begin{aligned} S_{0m} = 1\,; \qquad S_{i,m+1} &= S_{im} + m\, S_{i-1,m}, \qquad (i = 1, 2, \ldots, m),\\ \text{where}\quad S_{mm} &= 0\,. \end{aligned}\right\} \tag{2}$$

7. In the bounding-formula (4.11.38) of the remainder $R(u)$ of the Newton-Cotes formula, the following constants

$$b_m = \begin{cases} \dfrac{1}{m!} \displaystyle\int_{-1}^{1} \omega_m(x)\, dx & \text{(for } m \text{ even)},\\[2ex] \dfrac{1}{(m+1)!} \displaystyle\int_{-1}^{1} x\, \omega_m(x)\, dx & \text{(for } m \text{ odd)}, \end{cases}$$

appear. Show that, putting

$$c_m = \frac{1}{m!} \int_0^1 \prod_{i=1}^{m} (t - i + 1)\, dt\,, \tag{1}$$

we have $\left(\text{with } h_m = \frac{2}{m-1}\right)$:

$$b_m = \begin{cases} -\,2\, h_m^{m+1}\, c_{m+1} & \text{(for } m \text{ even)},\\ -\,h_m^{m+2}\,(2\, c_{m+2} + c_{m+1}) & \text{(for } m \text{ odd)}. \end{cases} \tag{2}$$

8. Keeping into account the well-known inequality $\Gamma(x + \alpha) \leqq x^\alpha\, \Gamma(x)$, $(x > 0, 0 \leqq \alpha \leqq 1)$ [see § 3.10, Problem 8], deduce from (2) of the above

problem that

$$|b_m| < \begin{cases} \dfrac{2\,h_m^{m+1}}{m\,(\log m)^2} & \text{(for } m \text{ even)}\,, \\ \dfrac{h_m^{m+2}}{m\,(\log m)^2} & \text{(for } m \text{ odd)}\,. \end{cases} \tag{1}$$

9. Compute the values of the constants b_m considered in the Problem 7, for $m = 2, 3, 4, 5, 6, 7$.

10. Show that, for the remainder $R(u)$ of the Tchebyschef quadrature formula (4.6.25) or (4.12.1) the following bounding formula

$$|R(u)| \leqq \pi \frac{(2\,p+1)!!}{[(p+1)!]^2}\,\omega_{p+1}\,, \qquad (p = 0, 1, \ldots, 2\,m-2) \tag{1}$$

holds, where ω_{p+1} means the *oscillation* of $u^{(p+1)}(x)$ in $[-1, 1]$.

11. Prove that all the coefficients A_{0i} of (4.13.4), written for $s = 0$, are positive.

12. Prove that the coefficients A_{0i} of the quadrature formula (4.13.4) fulfill the following inequality

$$\int_a^{x_\nu} p(x)\,dx < \sum_{i=1}^{\nu} A_{0i} < \int_a^{x_{\nu+1}} p(x)\,dx\,, \qquad (\nu = 1, 2, \ldots, m-1)\,. \tag{1}$$

Note that, considering (4.13.4) for $s = 0$, the inequality (1) can be used for the coefficients of the (4.5.2), (4.9.1), (4.10.1).

13. Assuming that $[a, b]$ be finite, consider the problem of the construction of a quadrature formula of the following type [see (4.13.1)]:

$$\int_a^b p(x)\,u(x)\,dx = \sum_{h=0}^{r} \sum_{i=0}^{m+1} A_{hi}\,u^{(h)}(x_i) + R(u)\,, \tag{1}$$

where

$$a = x_0 < x_1 < \cdots < x_m < x_{m+1} = b\,, \tag{2}$$

with the condition $R(u) = 0$ where $u(x)$ is a polynomial of degree $\leqq m\,(r+2) + 2\,r + 1$.

CHAPTER 5

Generalized quadrature formulae and questions of convergence

5.1. Generalized quadrature formulae in the case of the constant weight.[1]) In Chapter 2 we stated the concept of *elementary* quadrature formula:

$$\int_a^b g(x)\, u(x)\, dx = \sum_{h=0}^{n-1} \sum_{i=1}^{m} A_{hi}\, u^{(h)}(x_i) + R(u)\,, \quad E(u) = 0 \Rightarrow R(u) = 0\,. \quad (5.1.1)$$

Such a formula does not always yields for the integral $\int_a^b g(x)\, u(x)\, dx$ a value sufficiently approximated and therefore, if we want to evaluate the integral with a *prefixed* approximation, we need to operate differently from the simple application of (5.1.1). In the case that the interval $[a, b]$ be *finite*,[2]) the most used procedure is that of dividing $[a, b]$ into a certain number ν of partial intervals $[\alpha_{j-1}, \alpha_j]$, $(j = 1, 2, \ldots, \nu)$ defined by the points

$$a = \alpha_0 < \alpha_1 < \alpha_2 < \cdots < \alpha_{\nu-1} < \alpha_\nu = b\,, \quad (5.1.2)$$

and successively of applying the formula (5.1.1) to each of the integrals $\int_{\alpha_{j-1}}^{\alpha_j} g(x)\, u(x)\, dx$, of course after having performed a linear change of the variable which lead the interval $[\alpha_{j-1}, \alpha_j]$ on the original interval $[a, b]$. Such a change of variable generally modifies both the argument function $u(x)$ and *the weight function* $g(x)$; the changing of $u(x)$ is not important, but it is clear that, if the weight $g(x)$ changes, (5.1.1) is no more applicable. Therefore the above procedure is valid only if, changing the variable, the weight $g(x)$ remains unchanged, *that is only if* $g(x) =$ constant.

We shall therefore begin by studying this case $g(x) \equiv 1$, however observing that, by means of an opportune device (see next § 5.2), the method can be extended to the case of any weight $g(x)$. Let us therefore start from an elementary quadrature formula of the type

$$\int_a^b u(x)\, dx = \sum_{h=0}^{n-1} \sum_{i=1}^{m} A_{hi}\, u^{(h)}(x_i) + R(u)\,, \qquad E(u) = 0 \Rightarrow R(u) = 0\,, \quad (5.1.3)$$

and, divided $[a, b]$ into ν partial intervals, by means of the points shown in

[1]) See A. Ghizzetti [2].

[2]) If $[a, b]$ is *infinite* we may modify the integral by a change of variable which lead the infinite interval on a finite one. We may also opportunely extend the procedure that we shall present in this § and in the next one, referring to *two* elementary quadrature formulae: one on finite interval and another on infinite interval.

[5.1.2], we write, putting $\delta_j = \alpha_j - \alpha_{j-1}$:

$$\int_a^b u(x)\,dx = \sum_{j=1}^{\nu} \int_{\alpha_{j-1}}^{\alpha_j} u(\xi)\,d\xi = \sum_{j=1}^{\nu} \frac{\delta_j}{b-a} \int_a^b u\left(\alpha_{j-1} + \delta_j \frac{x-a}{b-a}\right) dx$$

$$= \sum_{j=1}^{\nu} \frac{\delta_j}{b-a} \left\{ \sum_{h=0}^{n-1} \sum_{i=1}^{m} A_{hi} \left[\frac{d^h}{dx^h} u\left(\alpha_{j-1} + \delta_j \frac{x-a}{b-a}\right)\right]_{x=x_i} + R\left[u\left(\alpha_{j-1} + \delta_j \frac{x-a}{b-a}\right)\right]\right\},$$

and thus we obtain the formula

$$\int_a^b u(x)\,dx = \sum_{j=1}^{\nu} \sum_{h=0}^{n-1} \sum_{i=1}^{m} \left(\frac{\delta_j}{b-a}\right)^{h+1} A_{hi}\, u^{(h)}\left(\alpha_{j-1} + \delta_j \frac{x_i - a}{b-a}\right) + \varrho(u)\,, \quad (5.1.4)$$

with

$$\varrho(u) = \sum_{j=1}^{\nu} \frac{\delta_j}{b-a} R\left[u\left(\alpha_{j-1} + \delta_j \frac{x-a}{b-a}\right)\right],$$

that we call *generalized quadrature formula* connected to the elementary formula (5.1.3).

As to the remainder $\varrho(u)$ of (5.1.4), recalling the expression of $R(u)$ given by (2.1.13) and the one of the operator E given by (1.1.1), we easy find

$$\varrho(u) = \sum_{j=1}^{\nu} \sum_{k=0}^{n} \sum_{i=0}^{m} \left(\frac{\delta_j}{b-a}\right)^{n-k+1} \int_{x_i}^{x_{i+1}} \varphi_i(x)\, a_k(x)\, u^{(n-k)}\left(\alpha_{j-1} + \delta_j \frac{x-a}{b-a}\right) dx\,. \quad (5.1.5)$$

Having thus stated (5.1.4), (5.1.5), we must consider the problem whether or not they define a *convergent procedure of calculus*, that is if, putting $\delta = \max_{j=1,\ldots,n} \delta_j$, there results or not

$$\lim_{\delta \to 0} \varrho(u) = 0\,. \quad (5.1.6)$$

To this aim we prove the following theorem:

Theorem 5.1.I. *For every $u(x) \in A\, C^{n-1}[a, b]$, the remainder $\varrho(u)$ of the generalized quadrature formula (5.1.4) has the property expressed by*

$$\lim_{\delta \to 0} \varrho(u) = \frac{1}{b-a} \int_a^b u(x)\,dx \cdot R(1)\,.\,[1] \quad (5.1.7)$$

[1]) Remember that, on the base of (2.1.13) and (1.1.1), we have

$$R(1) = \sum_{i=0}^{m} \int_{x_i}^{x_{i+1}} \varphi_i(x)\, E(1)\, dx = \sum_{i=0}^{m} \int_{x_i}^{x_{i+1}} \varphi_i(x)\, a_n(x)\, dx;$$

instead of $R(1)$ we may also write $\varrho(1)$ because from (5.1.5) there follows

$$\varrho(1) = \sum_{j=1}^{\nu} \frac{\delta_j}{b-a} \cdot \sum_{i=0}^{m} \int_{x_i}^{x_{i+1}} \varphi_i(x)\, a_n(x)\, dx \quad \text{and we have} \quad \sum_{i=1}^{\nu} \frac{\delta_j}{b-a} = 1.$$

Proof. Starting form (5.1.5), we write $\varrho(u) = \varrho_0(u) + \varrho_1(u)$ with

$$\varrho_0(u) = \sum_{j=1}^{\nu} \sum_{k=0}^{n-1} \sum_{i=0}^{m} \left(\frac{\delta_j}{b-a}\right)^{n-k+1} \int_{x_i}^{x_{i+1}} \varphi_i(x)\, a_k(x)\, u^{(n-k)}\left(\alpha_{j-1} + \delta_j \frac{x-a}{b-a}\right) dx\,, \tag{5.1.8}$$

$$\varrho_1(u) = \sum_{j=1}^{\nu} \sum_{i=0}^{m}{}' \frac{\delta_j}{b-a} \int_{x_i}^{x_{i+1}} \varphi_i(x)\, a_n(x)\, u\left(\alpha_{j-1} + \delta_j \frac{x-a}{b-a}\right) dx\,. \tag{5.1.9}$$

Having put

$$\max_{\substack{a \leqq x \leqq b \\ i=0,1,\ldots,m}} |\varphi_i(x)| = H\,; \quad \max_{a \leqq x \leqq b} |a_k(x)| = A_k\,, \quad (k = 0, 1, \ldots, n-1)\,, \tag{5.1.10}$$

from (5.1.8) we obtain

$$|\varrho_0(u)| \leqq H \sum_{j=1}^{\nu} \sum_{k=0}^{n-1} A_k \left(\frac{\delta_j}{b-a}\right)^{n-k+1} \sum_{i=0}^{m} \int_{x_i}^{x_{i+1}} \left|u^{(n-k)}\left(\alpha_{j-1} + \delta_j \frac{x-a}{b-a}\right)\right| dx$$

$$= H \sum_{j=1}^{\nu} \sum_{k=0}^{n-1} A_k \left(\frac{\delta_j}{b-a}\right)^{n-k+1} \int_a^b \left|u^{(n-k)}\left(\alpha_{j-1} + \delta_j \frac{x-a}{b-a}\right)\right| dx$$

$$= H \sum_{j=1}^{\nu} \sum_{k=0}^{n-1} A_k \left(\frac{\delta_j}{b-a}\right)^{n-k} \int_{\alpha_{j-1}}^{\alpha_j} |u^{(n-k)}(\xi)|\, d\xi \leqq H \sum_{k=0}^{n-1} A_k \left(\frac{\delta}{b-a}\right)^{n-k}$$

$$\times \sum_{j=1}^{\nu} \int_{\alpha_{j-1}}^{\alpha_j} |u^{(n-k)}(\xi)|\, d\xi = H \frac{\delta}{b-a} \sum_{k=0}^{n-1} A_k \left(\frac{\delta}{b-a}\right)^{n-k-1} \int_a^b |u^{(n-k)}(\xi)|\, d\xi\,,$$

and therefore we may affirm that

$$\lim_{\delta \to 0} \varrho_0(u) = 0\,. \tag{5.1.11}$$

From (5.1.9) we then get

$$\varrho_1(u) = \frac{1}{b-a} \sum_{i=0}^{m} \int_{x_i}^{x_{i+1}} \varphi_i(x)\, a_n(x) \sum_{j=1}^{\nu} \delta_j\, u\left(\alpha_{j-1} + \delta_j \frac{x-a}{b-a}\right) dx$$

and since we have

$$\lim_{\delta \to 0} \sum_{j=1}^{\nu} \delta_j\, u\left(\alpha_{j-1} + \delta_j \frac{x-a}{b-a}\right) = \int_a^b u(x)\, dx \quad \text{(uniformly for } a \leqq x \leqq b)\,,$$

we deduce that

$$\lim_{\delta \to 0} \varrho_1(u) = \frac{1}{b-a} \int_a^b u(x)\, dx \sum_{i=0}^{m} \int_{x_i}^{x_{i+1}} \varphi_i(x)\, a_n(x)\, dx$$

$$= \frac{1}{b-a} \int_a^b u(x)\, dx \cdot R(1)\,. \tag{5.1.12}$$

From (5.1.11) and (5.1.12) there follows (5.1.7), q.e.d.

From the theorem now proved there follow these others:

Theorem 5.1.II. *Necessary and sufficient condition for there results* $\lim_{\delta\to 0} \varrho(u) = 0$, *with any* $u \in A\ C^{n-1}[a, b]$, *is that* $R(1) = 0$, *that is the elementary quadrature formula (5.1.3) must be exact for the function* $u(x) \equiv 1$.[1]) *This condition can be expressed also in the form*

$$\sum_{i=1}^{m} A_{0i} = b - a\,. \tag{5.1.13}$$

Theorem 5.1.III. *If* $a_n(x) \equiv 0$, *that is if in the prefixed linear differential operator* E *does not occur the term of order zero, so we may put*

$$E = \sum_{k=0}^{n-p} a_k(x) \frac{d^{n-k}}{dx^{n-k}} \quad [\text{with } 1 \leqq p \leqq n\,; \quad a_{n-p}(x) \not\equiv 0]\,, \tag{5.1.14}$$

then we surely have $\lim_{\delta\to 0} \varrho(u) = 0$ *and more precisely*

$$\varrho(u) = o(\delta^p)\,, \quad (\text{for } \delta \to 0)\,, \tag{5.1.15}$$

with any $u(x) \in A\ C^{n-1}[a, b]$.

Proof. The condition $a_n(x) \equiv 0$ implies $E(1) = 0$ and therefore $R(1) = 0$, whence for the preceding theorem we have $\lim_{\delta\to 0} \varrho(u) = 0$. To prove (5.1.15) we observe that, if (5.1.14) holds, we have $\varrho_1(u) = 0$ [see (5.1.9)] and $A_k = 0$ for $k > n - p$ [see (5.1.10)]. Then, considering again the bound obtained for $|\varrho_0(u)|$ in the proof of theorem 5.1.I, we may write

$$|\varrho(u)| = |\varrho_0(u)| \leqq H \frac{\delta}{b-a} \sum_{k=0}^{n-p} A_k \left(\frac{\delta}{b-a}\right)^{n-k-1} \int_a^b |u^{(n-k)}(\xi)|\, d\xi$$

$$= \left(\frac{\delta}{b-a}\right)^p \cdot H \sum_{k=0}^{n-p} A_k \left(\frac{\delta}{b-a}\right)^{n-p-k} \int_a^b |u^{(n-k)}(\xi)|\, d\xi$$

and from here (5.1.15) immediately follows, q.e.d.

We now have only to examine the case where (5.1.13) is not satisfied and therefore there does not result $\lim_{\delta\to 0} \varrho(u) = 0$, so that the generalized quadrature formula (5.1.4) does not yield a convergent procedure of calculus. Having put

$$S_0 = \sum_{i=1}^{m} A_{0i} \quad \text{or} \quad S_0 = b - a - R(1)\,,\ ^{2)} \tag{5.1.16}$$

[1]) We may also say that the *generalized quadrature formula* (5.1.4) *be exact for the function* $u(x) \equiv 1$; in fact we have seen that $\varrho(1) = R(1)$.

[2]) This second expression of S_0 follows from (5.1.3) putting $u(x) = 1$.

let us assume then that it be $S_0 \neq b - a$ and point out two subcases according to which there results $S_0 \neq 0$ or $S_0 = 0$. *In the first subcase $S_0 \neq b - a$ and $S_0 \neq 0$* we have for (5.1.7)

$$\lim_{\delta \to 0} \varrho(u) = \frac{1}{b-a} \int_a^b u(x)\,dx \cdot (b - a - S_0) = \left(1 - \frac{S_0}{b-a}\right) \int_a^b u(x)\,dx$$

and therefore we may modify the generalized formula (5.1.4) in the following way:

$$\int_a^b u(x)\,dx - \left(1 - \frac{S_0}{b-a}\right) \int_a^b u(x)\,dx$$

$$= \sum_{j=1}^{\nu} \sum_{h=0}^{n-1} \sum_{i=1}^{m} \left(\frac{\delta_j}{b-a}\right)^{h+1} A_{hi}\, u^{(h)}\left(\alpha_{j-1} + \delta_j \frac{x_i - a}{b-a}\right) + \left[\varrho(u) - \lim_{\delta \to 0} \varrho(u)\right].$$

We deduce from it

$$\int_a^b u(x)\,dx = \frac{b-a}{S_0} \sum_{j=1}^{\nu} \sum_{h=0}^{n-1} \sum_{i=1}^{m} \left(\frac{\delta_j}{b-a}\right)^{h+1} A_{hi}\, u^{(h)}\left(\alpha_{j-1} + \delta_j \frac{x_i - a}{b-a}\right) + \bar{\varrho}(u) \tag{5.1.17}$$

with

$$\bar{\varrho}(u) = \frac{b-a}{S_0} [\varrho(u) - \lim_{\delta \to 0} \varrho(u)], \quad \lim_{\delta \to 0} \bar{\varrho}(u) = 0,$$

whence *the modified formula (5.1.17) yields again a convergent procedure of calculus.*

In the second subcase $S_0 = 0$ we have $\lim_{\delta \to 0} \varrho(u) = \int_a^b u(x)\,dx$ and therefore (5.1.4) is of no use since the remainder $\varrho(u)$ tends to the integral itself which must be computed, while the approximating part tends to zero; that is there occurs a situation opposite to the one which appeared in the preceding cases. *The occurrence $S_0 = 0$ represents then the only irremediable case of non convergence*; using (2.1.12) and (5.1.16) this case occurs when

$$\sum_{i=1}^{m} [E^*_{n-1}(\varphi_i - \varphi_{i-1})]_{x=x_i} = 0$$

and it is clear that it can be avoided using the arbitrarity of the nodes and of the solutions $\varphi_1(x), \ldots, \varphi_{m-1}(x)$ of the differential equation $E^*(\varphi) = 1$.

We now add some remarks of the preceding results.

Remark 1. Theorem 5.1.III shows the opportunity of choosing differential operators E lacking of the term of order 0; moreover it shows that the infinitesimal order of the remainder $\varrho(u)$ coincides (in general) with the minimum order p of the derivations occurring in E. Therefore it is convenient to choose

p as high as possible and from *this point of view the classic formulae which correspond to* $E = \frac{d^n}{dx^n}$ *(that is to* $p = n$*) are privileged.*

Remark 2. The elementary formula (5.1.3) is exact when $E(u) = 0$. *This is no more true in general for the generalized formula (5.1.4),* since this was obtained by applying the elementary formula to the various function $u\left(\alpha_{j-1} + \delta_j \frac{x-a}{b-a}\right)$ and the differential equation $E(u) = 0$ is not in general invariant with respect to the changes of variable $\xi = \alpha_{j-1} + \delta_j \frac{x-a}{b-a}$.

Evidently the classic case $E = \frac{d^n}{dx^n}$ *represents an exception* and, in this case, (5.1.5) becomes

$$\varrho(u) = \sum_{j=1}^{\nu} \sum_{i=0}^{m} \left(\frac{\delta_j}{b-a}\right)^{n+1} \int_{x_i}^{x_{i+1}} \varphi_i(x)\, u^{(n)}\left(\alpha_{j-1} + \delta_j \frac{x-a}{b-a}\right) dx\,, \quad (5.1.18)$$

whence the generalized formula is, like the elementary one, exact for the polynomials of degree $\leq n-1$.

Remark 3. The previous considerations can induce to think that it is sufficient to limit oneself to the classic formulae and that it is useless to consider the quadrature formulae relative to more general operators. Howewer this is not true, since what is important in practice is only to evaluate the order of magnitude of the remainder $\varrho(u)$ *for the particular* $u(x)$ *which is to be integrated.* Now it may happen that for such $u(x)$ it is practically impossible to obtain an evaluation of the remainder (5.1.18), on the contrary it can be easier using the expression (5.1.5), when we choose conveniently the coefficients $a_k(x)$ of the differential operator E. In other words the general method that we presented allows to try *case by case* to construct a quadrature formula which yields an easy evaluation of the remainder.

5.2. Generalized quadrature formulae with any weight.[1]) Let us consider again the formula

$$\int_a^b g(x)\, u(x)\, dx = \sum_{h=0}^{n-1} \sum_{i=1}^{m} A_{hi}\, u^{(h)}(x_i) + R(u)\;; \quad E(u) = 0 \Rightarrow R(u) = 0\,, \quad (5.2.1)$$

assuming $[a, b]$ *finite* and $g(x)$ non constant. We recall that to (5.2.1) there are connected certain determined solutions of the differential equation $E^*(\varphi) = g(x)$ and precisely the two following ones [see (2.1.9)]:

$$\varphi_0(x) = -\int_a^x K(\xi, x)\, g(\xi)\, d\xi\,, \qquad \varphi_m(x) = \int_x^b K(\xi, x)\, g(\xi)\, d\xi\,, \qquad (5.2.2)$$

[1]) See A. Ghizzetti [2].

together with others $m - 1$ that it is convenient to write here in the form

$$\varphi_i(x) = \frac{1}{2}[\varphi_0(x) + \varphi_m(x)] + \sum_{r=1}^{n} c_{ir} v_r(x), \qquad (i = 1, 2, \ldots, m-1), \quad (5.2.3)$$

where c_{ir} are constants and $[v_1(x), \ldots, v_n(x)]$ is a fundamental system of solutions of $E^*(v) = 0$.

Now, if we operate on (5.2.1) a computation analogous to that which led us from the elementary formula (5.1.3) to the generalized formula (5.1.4), it is clear that, proceeding as follows

$$\int_a^b g(x)\, u(x)\, dx = \sum_{j=1}^{\nu} \int_{\alpha_{j-1}}^{\alpha_j} g(\xi)\, u(\xi)\, d\xi$$

$$= \sum_{j=1}^{\nu} \frac{\delta_j}{b-a} \int_a^b g\left(\alpha_{j-1} + \delta_j \frac{x-a}{b-a}\right) u\left(\alpha_{j-1} + \delta_j \frac{x-a}{b-a}\right) dx, \qquad (5.2.4)$$

we cannot any more apply to these last integrals the elementary formula (5.2.1) since in them there occurs a weight different from $g(x)$. The difficulty can be overcome in the following way. Putting

$$\lambda_j = \frac{\delta_j}{b-a}, \qquad \mu_j = \frac{(b-a)\,\alpha_{j-1} - a\,\delta_j}{b-a-\delta_j}, \qquad (j = 1, 2, \ldots, \nu), \quad (5.2.5)$$

there results

$$0 < \lambda_j < 1, \quad a \leqq \mu_j \leqq b\,,$$ [1]

$$\alpha_{j-1} + \delta_j \frac{x-a}{b-a} = \lambda_j x + (1 - \lambda_j)\,\mu_j\,,$$

so that it would be possible to pass from (5.2.4) to a generalized formula in the case we could enjoy an elementary formula for the integral

$$\int_a^b g\,[\lambda x + (1-\lambda)\,\mu]\, u(x)\, dx\,, \qquad (5.2.6)$$

where there occurs a weight depending on two parameters λ, μ variable in the set

$$T\,(0 < \lambda \leqq 1\,;\; a \leqq \mu \leqq b)\,. \qquad (5.2.7)$$ [2]

But such a formula can be immediately constructed by the rule of § 2.1.

Precisely, having considered the differential equation

$$E^*(\varphi) = g\,[\lambda x + (1-\lambda)\,\mu]$$

[1]) We assume $\nu > 1$; we may observe that we have more precisely

$$a = \mu_1 < \mu_2 < \cdots < \mu_\nu = b\,.$$

[2]) We have included the value $\lambda = 1$ so that the integral $\int_a^b g(x)\, u(x)\, dx$, which occurs in (5.2.1), may be considered as a particular case of the integral (5.2.6).

and of it the following solutions [see (5.2.2) and (5.2.3)]:

$$\left.\begin{aligned}\varphi_0(x;\lambda,\mu) &= -\int_a^x K(\xi,x)\, g\,[\lambda\,\xi + (1-\lambda)\,\mu]\, d\xi\,,\\ \varphi_m(x;\lambda,\mu) &= \int_x^b K(\xi,x)\, g\,[\lambda\,\xi + (1-\lambda)\,\mu]\, d\xi\,,\end{aligned}\right\} \tag{5.2.8}$$

$$\left.\begin{aligned}\varphi_i(x;\lambda,\mu) &= \frac{1}{2}\,[\varphi_0(x;\lambda,\mu) + \varphi_m(x;\lambda,\mu)] + \sum_{r=1}^{n} c_{ir}(\lambda,\mu)\, v_r(x)\,,\\ &(i = 1, 2, \ldots, m-1)\,,\end{aligned}\right\} \tag{5.2.9}$$

where the functions $c_{ir}(\lambda, \mu)$ for the moment can be fixed arbitrarily, we get the elementary formula [see (5.2.1)]:

$$\int_a^b g\,[\lambda\,x + (1-\lambda)\,\mu]\, u(x)\, dx = \sum_{h=0}^{n-1} \sum_{i=1}^{m} A_{hi}(\lambda,\mu)\, u^{(h)}(x_i) + R(u;\lambda,\mu)\,, \tag{5.2.10}$$

with [see (2.1.12) and (2.1.13)]:

$$A_{hi}(\lambda,\mu) = \{E^*_{n-h-1}[\varphi_i(x;\lambda,\mu) - \varphi_{i-1}(x;\lambda,\mu)]\}_{x=x_i}\,, \tag{5.2.11}$$

$$R(u;\lambda,\mu) = \sum_{i=0}^{m} \int_{x_i}^{x_{i+1}} \varphi_i(x;\lambda,\mu)\, E(u)\, dx\,. \tag{5.2.12}$$

We intend to consider shortly the way of choosing the functions $c_{ir}(\lambda, \mu)$, but now we write again (5.2.4) and we apply to each of the integrals

$$\int_a^b g\left(\alpha_{j-1} + \delta_j \frac{x-a}{b-a}\right) u\left(\alpha_{j-1} + \delta_j \frac{x-a}{b-a}\right) dx$$

$$= \int_a^b g\,[\lambda_j\, x + (1-\lambda_j)\,\mu_j]\, u\,[\lambda_j\, x + (1-\lambda_j)\,\mu_j]\, dx\,,$$

the formula (5.2.10), thus obtaining

$$\int_a^b g(x)\, u(x)\, dx = \sum_{j=1}^{\nu} \lambda_j \left(\sum_{h=0}^{n-1} \sum_{i=1}^{m} A_{hi}(\lambda_j,\mu_j) \left\{\frac{d^h}{dx^h} u\,[\lambda_j\, x + (1-\lambda_j)\,\mu_j]\right\}_{x=x_i}\right.$$

$$\left. + R\,\{u\,[\lambda_j\, x + (1-\lambda_j)\,\mu_j];\lambda_j,\mu_j\}\right).$$

We thus reached the *generalized formula*

$$\int_a^b g(x)\, u(x)\, dx = \sum_{j=1}^{\nu} \sum_{h=0}^{n-1} \sum_{i=1}^{m} \lambda_j^{h+1} A_{hi}(\lambda_j,\mu_j)\, u^{(h)}\,[\lambda_j\, x_i + (1-\lambda_j)\,\mu_j] + \varrho(u) \tag{5.2.13}$$

where the remainder $\varrho(u)$ is expressed by

$$\varrho(u) = \sum_{j=1}^{\nu} \lambda_j\, R\,\{u\,[\lambda_j\, x + (1-\lambda_j)\,\mu_j];\lambda_j,\mu_j\}\,.$$

that is for (5.2.12) and (1.1.1):

$$\varrho(u) = \sum_{j=1}^{\nu} \sum_{k=0}^{n} \sum_{i=0}^{m} \lambda_j^{n-k+1} \int_{x_i}^{x_{i+1}} \varphi_i(x;\lambda_j,\mu_j)\, a_k(x)\, u^{(n-k)}\,[\lambda_j\, x + (1-\lambda_j)\,\mu_j]\, dx\,. \tag{5.2.14}$$

Now it is convenient to put some limitations to the choice of the functions $c_{ir}(\lambda, \mu)$ which occur in (5.2.9). To this aim let us begin by recalling that we wanted to construct a generalized formula which would be connected to the elementary formula (5.2.1); on the contrary (5.2.13) appears connected to (5.2.10). Therefore it is to be desidered that there it be a close tie between the two elementary formulae (5.2.1) and (5.2.10).

If we compare their first members, we may observe that we have

$$\lim_{\lambda\to 1} \int_a^b g\,[\lambda\, x + (1-\lambda)\,\mu]\, u(x)\, dx = \int_a^b g(x)\, u(x)\, dx \tag{5.2.15}$$

uniformly with respect to $\mu \in [a, b]$.[1]) Therefore it is convenient to construct (5.2.10) in such a way that an analogous relation holds for each pair of terms corresponding in the second members of (5.2.1) and (5.2.10), that is in such a way that it be

$$\lim_{\lambda\to 1} A_{hi}(\lambda,\mu) = A_{hi}\,, \quad (0 \leqq h \leqq n-1,\ 1 \leqq i \leqq m)\ ; \quad \lim_{\lambda\to 1} R(u;\lambda,\mu) = R(u)\,, \tag{5.2.16}$$

uniformly with respect to $\mu \in [a, b]$.

To this aim we prove the following theorem:

Theorem 5.2.I. *If the function* $c_{ir}(\lambda, \mu)$, $(i = 1, 2, \ldots, m-1; r = 1, 2, \ldots, n)$ *satisfy the following hypotheses*

$$c_{ir}(\lambda,\mu)\ \ \textit{continuous}\ \text{in}\ T\ ; \quad c_{ir}(1,\mu) = c_{ir}\,, \tag{5.2.17}$$

where c_{ir} *are the constants which occur in (5.2.3), then (5.2.16) is valid and therefore the elementary formula (5.2.10) is to be thought obtained from the primitive formula (5.2.1) by means of a continuous variation of its terms.*

Proof. It is sufficient to prove that we have

$$\lim_{\lambda\to 1} \varphi_i^{(h)}(x;\lambda,\mu) = \varphi_i^{(h)}(x)\,, \quad (i = 0, 1, \ldots, m;\ h = 0, 1, \ldots, n-1)\,, \tag{5.2.18}$$

uniformly with respect to $x \in [a, b]$ *and* $\mu \in [a; b]$, since (5.2.16) immediately follows from (5.2.11), (5.2.12) and from the expressions of A_{hi} and $R(u)$

[1]) In fact, considering that $u(x)$ is continuous in $[a, b]$, the integral $\int_a^b g\,[\lambda\, x + (1-\lambda)\,\mu] \times u(x)\, dx = \frac{1}{\lambda} \int_{\lambda a+(1-\lambda)\mu}^{\lambda b+(1-\lambda)\mu} g(t)\, u\left(\frac{t-(1-\lambda)\,\mu}{\lambda}\right) dt$ is a continuous function of (λ, μ) in T and therefore uniformly continuous for instance for $\frac{1}{2} \leqq \lambda \leqq 1$, $a \leqq \mu \leqq b$.

given by (2.1.12), (2.1.13). From (5.2.8) there follows

$$\varphi_0^{(h)}(x;\lambda,\mu) = -\int_a^x \frac{\partial^h}{\partial x^h} K(\xi, x)\, g\,[\lambda\,\xi + (1-\lambda)\,\mu]\, d\xi$$

$$= -\frac{1}{\lambda}\int_{\lambda a+(1-\lambda)\mu}^{\lambda x+(1-\lambda)\mu} \frac{\varepsilon^h}{\partial x^h} K\left(\frac{t-(1-\lambda)\,\mu}{\lambda}, x\right) g(t)\, dt$$

and therefore, owing to the continuity of $\frac{\partial^h}{\partial x^h} K(\xi, x)$, we see that $\varphi_0^{(h)}(x;\lambda,\mu)$ is a continuous function for $x \in [a, b]$, $(\lambda, \mu) \in T$ and uniformly continuous for $x \in [a, b]$, $\lambda \in \left[\frac{1}{2}, 1\right]$, $\mu \in [a, b]$. There follows that, uniformly with respect to (x, μ), we have

$$\lim_{\lambda\to 1} \varphi_0^{(h)}(x;\lambda,\mu) = -\int_a^b \frac{\varepsilon^h}{\partial x^h} K(t, x)\, g(t)\, dt = \varphi_0^{(h)}(x)\,, \tag{5.2.19}$$

and analogously

$$\lim_{\lambda\to 1} \varphi_m^{(h)}(x;\lambda,\mu) = \varphi_m^{(h)}(x)\,. \tag{5.2.20}$$

For the remaining values of the index i, (5.2.18) soon follows from (5.2.9), using (5.2.17), (5.2.19), (5.2.20), (5.2.3). Theorem 5.2.I is thus proved.

However it is not sufficient to do on the functions $c_{ir}(\lambda, \mu)$ the hypotheses expressed by (5.2.17). In order to obtain that, at least in certain cases, the generalized formula (5.2.13) be convergent [in the sense that, putting $\delta = \max \delta_j$, we have $\lim_{\delta\to 0} \varrho(u)$] it is convenient to prescribe this other condition

$$\lim_{\lambda\to 0} \lambda\, c_{ir}(\lambda, \mu) = 0 \tag{5.2.21}$$

uniformly with respect to $\mu \in [a, b]$.[1]) We then have the following theorem:

Theorem 5.2.II. *Under the hypothesis (5.2.21) there results*

$$\lim_{\lambda\to 0} \lambda\, \varphi_i(x;\lambda,\mu) = 0\,, \qquad (i = 0, 1, \ldots, m)\,, \tag{5.2.22}$$

uniformly with respect to $x \in [a, b]$ *and* $\mu \in [a, b]$.

Proof. Putting $\max\limits_{\substack{a\le\xi\le b\\ a\le x\le b}} |K(\xi, x)| = M$, from (5.2.8) there follows

$$\left.\begin{array}{l} |\lambda\, \varphi_0(x;\lambda,\mu)| \\ |\lambda\, \varphi_m(x;\lambda,\mu)| \end{array}\right\} \le \lambda\, M \int_a^b |g\,[\lambda\xi + (1-\lambda)\,\mu]|\, d\xi = M \int_{\lambda a+(1-\lambda)\mu}^{\lambda b+(1-\lambda)\mu} |g(t)|\, dt\,. \tag{5.2.23}$$

For the absolute continuity of the integral of the function $g(t)$, summable in $[a, b]$, having fixed $\varepsilon > 0$, there exists a $\delta_\varepsilon > 0$ such that, for *every* interval

[1]) We suppose that $c_{ir}(\lambda, \mu)$, continuous in T $(0 < \lambda \le 1,\ a \le \mu \le b)$ may become infinite for $\lambda \to 0$, but of order lower that 1.

$[\alpha, \beta]$ contained in $[a, b]$ and of *length* $< \delta_\varepsilon$, we have $\int_\alpha^\beta |g(t)|\, dt < \varepsilon$. Then, assuming $\lambda(b - a) < \delta_\varepsilon$, from (5.2.23) we deduce $|\lambda\, \varphi_0(x; \lambda, \mu)|$, $|\lambda\, \varphi_m(x; \lambda, \mu)| < M\, \varepsilon$ and this proves (5.2.22) in the cases $i = 0$ and $i = m$.

Keeping into account this result, we pass to the other values of i starting from (5.2.9) and using the hypothesis (5.2.21).

Let us now study the remainder $\varrho(u)$ of the generalized formula (5.2.13). Under the only hypothesis of summability of the weight $g(x)$, it does not seem possible to state theorems analogous to theorems 5.1.I and 5.1.II which are valid in the case $g(x) \equiv 1$. However we may extend, under the hypothesis (5.2.21), theorem 5.1.III in the following way:

Theorem 5.2.III. *If $a_n(x) \equiv 0$, that is if in the prefixed linear differential operator E there does not occur the term of order zero, so that we can put*

$$E(u) = \sum_{k=0}^{n-p} a_k(x) \frac{d^{n-k}}{dx^{n-k}}, \qquad [\text{with } 1 \leqq p \leqq n,\ a_{n-p}(x) \not\equiv 0], \tag{5.2.24}$$

then we have $\lim_{\delta \to 0} \varrho(u) = 0$ *and more precisely*

$$\varrho(u) = o(\delta^{p-1}), \qquad (\text{for } \delta \to 0) \tag{5.2.25}$$

with any function $u(x) \in A\, C^{n-1}[a, b]$.

Proof. By virtue of (5.2.14) and (5.2.24) we may write

$$\varrho(u) = \sum_{j=1}^{\nu} \sum_{k=0}^{n-p} \sum_{i=0}^{m} \lambda_j^{n-k+1} \int_{x_i}^{x_{i+1}} \varphi_i(x; \lambda_j, \mu_j)\, a_k(x)\, u^{(n-k)}[\lambda_j\, x + (1 - \lambda_j)\, \mu_j]\, dx\,. \tag{5.2.26}$$

On the other hand for (5.2.22), having fixed $\varepsilon > 0$, there exists a $\delta_\varepsilon > 0$ such that for $0 < \lambda < \frac{\delta_\varepsilon}{b-a}$ we have $|\varphi_i(x; \lambda, \mu)| < \frac{\varepsilon}{\lambda}$ for $i = 0, 1, \ldots, m$ and with any x and μ in the interval $[a, b]$. Therefore, *assuming already* $\delta < \delta_\varepsilon$ $\left(\text{and therefore } \lambda_j = \frac{\delta_j}{b-a} \leqq \frac{\delta}{b-a} < \frac{\delta_\varepsilon}{b-a}\right)$ in (5.2.26) there results $|\varphi_i(x; \lambda_j, \mu_j)| < \frac{\varepsilon}{\lambda_j}$, so that, putting $A_k = \max_{a \leqq x \leqq b} |a_k(x)|$, we may write:

$$|\varrho(u)| < \sum_{j=1}^{\nu} \sum_{k=0}^{n-p} \sum_{i=0}^{m} \lambda_j^{n-k+1} \int_{x_i}^{x_{i+1}} \frac{\varepsilon}{\lambda_j} A_k\, |u^{(n-k)}[\lambda_j\, x + (1 - \lambda_j)\, \mu_j]|\, dx$$

$$= \varepsilon \sum_{j=1}^{\nu} \sum_{k=0}^{n-p} A_k\, \lambda_j^{n-k} \sum_{i=0}^{m} \int_{x_i}^{x_{i+1}} |u^{(n-k)}[\lambda_j\, x + (1 - \lambda_j)\, \mu_j]|\, dx$$

$$= \varepsilon \sum_{j=1}^{\nu} \sum_{k=0}^{n-p} A_k\, \lambda_j^{n-k} \int_a^b |u^{(n-k)}[\lambda_j\, x + (1 - \lambda_j)\, \mu_j]|\, dx$$

and successively, operating in this integral the substitution $\lambda_j x + (1 - \lambda_j)\mu_j = \xi$ and considering that we may also write $\xi = \alpha_{j-1} + \delta_j \frac{x-a}{b-a}$:

$$|\varrho(u)| < \varepsilon \sum_{j=1}^{\nu} \sum_{k=0}^{n-p} A_k \lambda_j^{n-k-1} \int_{\alpha_{j-1}}^{\alpha_j} |u^{(n-k)}(\xi)|\, d\xi$$

$$\leqq \varepsilon \sum_{j=1}^{\nu} \sum_{k=0}^{n-p} A_k \left(\frac{\delta}{b-a}\right)^{n-k-1} \int_{\alpha_{j-1}}^{\alpha_j} |u^{(n-k)}(\xi)|\, d\xi$$

$$= \varepsilon \sum_{k=0}^{n-p} A_k \left(\frac{\delta}{b-a}\right)^{n-k-1} \sum_{j=1}^{\nu} \int_{\alpha_{j-1}}^{\alpha_j} |u^{(n-k)}(\xi)|\, d\xi$$

$$= \varepsilon \left(\frac{\delta}{b-a}\right)^{p-1} \sum_{k=0}^{n-p} A_k \left(\frac{\delta}{b-a}\right)^{n-p-k} \int_a^b |u^{(n-k)}(\xi)|\, d\xi\,.$$

Obviously there follows (5.2.25), q.e.d.

Now we may repeat three remarks analogous to those done at the end of § 5.1, but it is convenient to add another one.

In the generalized formula (5.2.13) the coefficients A_{hi} depend on the values λ_j, μ_j taken by the parameters λ, μ in correspondence with the adopted subdivision into parts of the interval $[a, b]$. Generally such a dependence is not simple and can lead to complicated formulae, of no practical use. In such a case it is opportune to use (5.2.13) not referring to an *arbitrary law* of subdivision of the interval $[a, b]$, but referring to that *particular law* chosen with the intention of simplifying the more possible the expression of the coefficients A_{hi}. To this purpose see the example given in the next §.

5.3. An example of generalized quadrature formula. Starting from the data

$$a = -1\,, \quad b = 1\,, \quad g(x) = \frac{1}{\sqrt{1-x^2}}\,; \quad m = 2\,,$$

$$x_1 = -1\,, \quad x_2 = 1\,; \quad n = 2, \quad E = \frac{d^2}{dx^2}$$

to which there correspond (according to (5.2.2)):

$$\varphi_0(x) = x \arcsin x + \sqrt{1-x^2} + \frac{\pi}{2}x\,, \quad \varphi_2(x) = x \arcsin x + \sqrt{1-x^2} - \frac{\pi}{2}x\,,$$

and assuming

$$\varphi_1(x) = x \arcsin x + \sqrt{1-x^2} - \frac{\pi}{2}\,,$$

we soon find the following elementary quadrature formula [which is the analogous one of (5.2.1)]:

$$\int_{-1}^{1} \frac{u(x)}{\sqrt{1-x^2}}\,dx = \frac{\pi}{2}\,[u\,(-1) + u(1)] + R(u)\,, \qquad R(u) = \int_{-1}^{1} \varphi_1(x)\, u''(x)\, dx\,, \tag{5.3.1}$$

valid for every $u(x) \in A\,C^1\,[-1,1]$. Assuming $v_1(x) = 1 + x$ and $v_2(x) = 1 - x$, (5.2.3) can be written

$$\varphi_1(x) = \frac{1}{2}\,[\varphi_0(x) + \varphi_2(x)] - \frac{\pi}{4}\,v_1(x) - \frac{\pi}{4}\,v_2(x)$$

and therefore we have

$$c_{11} = -\frac{\pi}{4}\,, \quad c_{12} = -\frac{\pi}{4}\,. \tag{5.3.2}$$

Let us now construct the formula analogous to (5.2.10). Formulae (5.2.8), (5.2.9) give

$$\varphi_0(x;\lambda,\mu) = \int_{-1}^{x} \frac{x-\xi}{\sqrt{1-[\lambda\,\xi + (1-\lambda)\,\mu]^2}}\,d\xi\,,$$

$$\varphi_2(x;\lambda,\mu) = -\int_{x}^{1} \frac{x-\xi}{\sqrt{1-[\lambda\,\xi + (1-\lambda)\,\mu]^2}}\,d\xi\,, \text{ [1]}$$

$$\varphi_1(x;\lambda,\mu) = \frac{1}{2}\,[\varphi_0(x;\lambda,\mu) + \varphi_2(x;\lambda,\mu)] + c_{11}(\lambda,\mu)\,v_1(x) + c_{12}\,(\lambda,\mu)\,v_2(x)\,,$$

and if we want that in the quadrature formula we are constructing do not occur [as in (5.3.1)] the values $u'(-1)$ and $u'(1)$, we must choose $c_{11}(\lambda,\mu)$, $c_{12}(\lambda,\mu)$ in such a way that there results $\varphi_1(-1;\lambda,\mu) = \varphi_1(1;\lambda,\mu) = 0$. We find

$$\left.\begin{aligned} c_{11}(\lambda,\mu) &= -\frac{1}{4}\int_{-1}^{1} \frac{1-\xi}{\sqrt{1-[\lambda\,\xi + (1-\lambda)\,\mu]^2}}\,d\xi\,,\\ c_{12}(\lambda,\mu) &= -\frac{1}{4}\int_{-1}^{1} \frac{1+\xi}{\sqrt{1-[\lambda\,\xi + (1-\lambda)\,\mu]^2}}\,d\xi\,. \end{aligned}\right\} \tag{5.3.3}$$

After that, by means of simple computations, we get the wanted formula [see (5.2.10)]:

$$\int_{-1}^{1} \frac{u(x)}{\sqrt{1-[\lambda\,x + (1-\lambda)\,\mu]^2}}\,dx = A_{01}(\lambda,\mu)\,u(-1) + A_{02}(\lambda,\mu)\,u(1) + R(u;\lambda,\mu) \tag{5.3.4}$$

[1]) These integrals and the others which will follow can be computed in an elementary way, but for sake of brevity we shall compute them at the end of our study.

with [see (5.2.11) and (5.2.12)]:

$$\left.\begin{aligned} A_{01}(\lambda,\mu) &= -\varphi_1'(-1) = \frac{1}{2}\int_{-1}^{1}\frac{1-\xi}{\sqrt{1-[\lambda\,\xi+(1-\lambda)\,\mu]^2}}\,d\xi\,,\\ A_{02}(\lambda,\mu) &= \varphi_1'(1) = \frac{1}{2}\int_{-1}^{1}\frac{1+\xi}{\sqrt{1-[\lambda\,\xi+(1-\lambda)\,\mu]^2}}\,d\xi\,,\\ R(u;\lambda,\mu) &= \int_{-1}^{1}\varphi_1(x;\lambda,\mu)\,u''(x)\,dx\,. \end{aligned}\right\} \tag{5.3.5}$$

Before deducing from (5.3.4) the generalized formula analogous to (5.2.13) we remark that the hypotheses (5.2.17) and (5.2.21) are satisfied. The validity of (5.2.17) is pointed out by (5.3.2) and (5.3.3), since we have

$$c_{11}(1,\mu) = -\frac{1}{4}\int_{-1}^{1}\frac{1-\xi}{\sqrt{1-\xi^2}}\,d\xi = -\frac{\pi}{4} = c_{11}\,,$$

$$c_{12}(1,\mu) = -\frac{1}{4}\int_{-1}^{1}\frac{1+\xi}{\sqrt{1-\xi^2}}\,d\xi = -\frac{\pi}{4} = c_{12}\,,$$

From formulae (5.3.3) there follows also that $|c_{11}(\lambda,\mu)|$ and $|c_{12}(\lambda,\mu)|$ do not exceed $\frac{1}{2}\int_{-1}^{1}\frac{d\xi}{\sqrt{1-[\lambda\,\xi+(1-\lambda)\,\mu]^2}}$ and therefore the validity of (5.2.21) will be assured if we show that

$$\lim_{\lambda\to 0}\lambda\int_{-1}^{1}\frac{d\xi}{\sqrt{1-[\lambda\,\xi+(1-\lambda)\,\mu]^2}}$$
$$= \lim_{\lambda\to 0}\{\arccos[-\lambda+(1-\lambda)\,\mu] - \arccos[\lambda+(1-\lambda)\,\mu]\} = 0$$

uniformly with respect to $\mu \in [-1, 1]$.

But this is evident because the function $\arccos[-\lambda+(1-\lambda)\,\mu] - \arccos[\lambda+(1-\lambda)\,\mu]$ is continuous (and therefore *uniformly continuous*) on the rectangle $[0 \leqq \lambda \leqq 1,\ -1 \leqq \mu \leqq 1]$ and has value 0 on the side $\lambda = 0$.

Then, with

$$-1 = \alpha_0 < \alpha_1 < \cdots < \alpha_{\nu-1} < \alpha_\nu = 1\,,\quad \delta_j = \alpha_j - \alpha_{j-1}\,,$$
$$\lambda_j = \frac{\delta_j}{2}\,,\quad \mu_j = \frac{2\,\alpha_{j-1}+\delta_j}{2-\delta_j}\,,$$

the generalized formula analogous to (5.2.13) is written

$$\int_{-1}^{1} \frac{u(x)}{\sqrt{1-x^2}}\,dx = \sum_{j=1}^{\nu} \lambda_j \,[A_{01}(\lambda_j, \mu_j)\, u(\alpha_{j-1}) + A_{02}(\lambda_j, \mu_j)\, u(\alpha_j)] + \varrho(u)$$

or better

$$\int_{-1}^{1} \frac{u(x)}{\sqrt{1-x^2}}\,dx = \lambda_1 A_{01}(\lambda_1, \mu_1)\, u(-1) + \sum_{j=1}^{\nu-1} [\lambda_{j+1} A_{01}(\lambda_{j+1}, \mu_{j+1})$$

$$+ \lambda_j A_{02}(\lambda_j, \mu_j)]\, u(\alpha_j) + \lambda_\nu A_{02}(\lambda_\nu, \mu_\nu)\, u(1) + \varrho(u)\,, \qquad (5.3.6)$$

and theorem 5.2.III assures us that, with $\delta = \max \delta_j$, we have

$$\varrho(u) = o(\delta) \qquad (\text{for } \delta \to 0)\,. \qquad (5.3.7)$$

In the formula (5.3.6) the coefficients $A_{01}(\lambda_j, \mu_j)$, $A_{02}(\lambda_j, \mu_j)$ written in connection with an *arbitrary* subdivision of $[a, b]$ into partial intervals, have expressions rather complicated and therefore the formula itself comes out to be of little practical utility. But, having remarked that by the substitution $\lambda\,\xi + (1-\lambda)\,\mu = \cos t$, $(0 \leqq t \leqq \pi)$, formulae (5.3.5) give[1])

$$A_{01}(\lambda_j, \mu_j) = \frac{1}{2\lambda_j^2} \int_{\operatorname{arcos} \alpha_j}^{\operatorname{arcos} \alpha_{j-1}} (\alpha_j - \cos t)\, dt\,,$$

$$A_{02}(\lambda_j, \mu_j) = \frac{1}{2\lambda_j^2} \int_{\operatorname{arcos} \alpha_j}^{\operatorname{arcos} \alpha_{j-1}} (\cos t - \alpha_{j-1})\, dt\,,$$

we see that these formulae can be greatly simplified if we assume for instance

$$\alpha_j = \cos\left(1 - \frac{j}{\nu}\right)\pi\,, \qquad (j = 0, 1, \ldots, \nu)\,. \qquad (5.3.8)$$

With such a choice, in fact, there results

$$\lambda_j A_{01}(\lambda_j, \mu_j) = \frac{1}{\cos\left(1-\frac{j}{\nu}\right)\pi - \cos\left(1 - \frac{j-1}{\nu}\pi\right)}$$

$$\times \int_{\left(1-\frac{j}{\nu}\right)\pi}^{\left(1-\frac{j-1}{\nu}\right)\pi} \left[\cos\left(1-\frac{j}{\nu}\right)\pi - \cos t\right] dt$$

$$= \frac{1}{2\sin\frac{2j-1}{2\nu}\pi \sin\frac{\pi}{2\nu}} \left(-\frac{\pi}{\nu}\cos\frac{j}{\nu}\pi + \sin\frac{j}{\nu}\pi - \sin\frac{j-1}{\nu}\pi\right)$$

[1]) Recall that $-\lambda_j + (1-\lambda_j)\,\mu_j = \alpha_{j-1}$, $\lambda_j + (1-\lambda_j)\,\mu_j = \alpha_j$.

$$= \frac{1}{2\sin\frac{2j-1}{2\nu}\pi\sin\frac{\pi}{2\nu}}\left(-\frac{\pi}{\nu}\cos\frac{2j-1}{2\nu}\pi\cos\frac{\pi}{2\nu}\right.$$

$$\left.+\frac{\pi}{\nu}\sin\frac{2j-1}{2\nu}\pi\sin\frac{\pi}{2\nu}+2\cos\frac{2j-1}{2\nu}\pi\sin\frac{\pi}{2\nu}\right)$$

$$=\frac{\pi}{2\nu}+\left(1-\frac{\pi}{2\nu}\operatorname{cotan}\frac{\pi}{2\nu}\right)\operatorname{cotan}\frac{2j-1}{2\nu}\pi$$

and analogously

$$\lambda_j A_{02}(\lambda_j,\mu_j)=\frac{\pi}{2\nu}-\left(1-\frac{\pi}{2\nu}\operatorname{cotan}\frac{\pi}{2\nu}\right)\operatorname{cotan}\frac{2j-1}{2\nu}\pi\,.$$

With that (5.3.6) becomes

$$\int_{-1}^{1}\frac{u(x)}{\sqrt{1-x^2}}\,dx=\frac{1}{2}\gamma_0\,u(-1)+\sum_{j=1}^{\nu-1}\gamma_j\,u\left(-\cos\frac{j}{\nu}\pi\right)+\frac{1}{2}\gamma_\nu\,u(-1)+\varrho(u) \tag{5.3.9}$$

where we put

$$\gamma_j=\frac{\pi}{\nu}-\left(1-\frac{\pi}{2\nu}\operatorname{cotan}\frac{\pi}{2\nu}\right)\left(\operatorname{cotan}\frac{2j-1}{2\nu}\pi-\operatorname{cotan}\frac{2j+1}{2\nu}\pi\right), \tag{5.3.10}$$

$$(\gamma_j=\gamma_{\nu-j};\qquad j=0,1,\ldots,\nu)\,.$$

It is soon seen that, with the α_j given by (5.3.8), the parameter δ comes out to be infinitesimal of the 1° order for $\nu\to+\infty$ and therefore (5.3.7) can be substituted by

$$\varrho(u)=o\left(\frac{1}{\nu}\right),\qquad(\text{for }\nu\to\infty)\,. \tag{5.3.11}$$

Formula (5.3.9) is exact when $u(x)$ is a polynomial of 1° degree.

5.4. Problems

1. Write the generalized quadrature formula corresponding to the elementary formula of the circumscribed trapezium [see (4.2.1)] and to a subdivision of the integration interval into ν equal parts. Give bounds for the remainder $\varrho(u)$.

2. Same task with reference to the elementary quadrature formula of the inscribed trapezium [see (4.2.6)].

3. Same problem with reference to the Cavalieri-Simpson elementary quadrature formula [see (4.2.11)].

4. Same problem with reference to the Euler-Mac Laurin elementary quadrature formula [see (4.3.3)].

5. Write a generalized quadrature formula corresponding to the elementary formula (4.4.19), choosing the subdivision of the integration interval with a device analogous to that used in the problem of § 5.3.

CHAPTER 6

Solutions to problems

6.1. Solutions to problems of § 1.5

1. The adjoint equations come out to be:

$$E^*(v) \equiv (-1)^n \frac{d^n v}{dx^n} = 0\,, \tag{1'}$$

$$E^*(v) \equiv \sum_{k=0}^{n} (-1)^{n-k}\, a_k\, v^{(n-k)} = 0\,, \tag{2'}$$

$$\left.\begin{aligned} &E^*(v) \equiv \sum_{k=0}^{n} \frac{a_k^*}{(n-k)!}\, \frac{v^{(n-k)}}{x^k} = 0\,,\\ &a_0^* = (-1)^n a_0; \quad a_k^* = (-1)^{n-k} \sum_{r=1}^{k} \binom{k-1}{r-1} a_r\,, \quad (k = 1, 2, \ldots, n)\,. \end{aligned}\right\} \tag{3'}$$

In the case (1) we may assume

$$u_i(x) = \frac{x^{i-1}}{(i-1)!}\,, \qquad v_i(x) = (-1)^{n-i} \frac{x^{n-i}}{(n-i)!}\,, \qquad (i = 1, 2, \ldots, n)\,.$$

It is sufficient to verify that (1.3.5) is satisfied; in fact for $h = 1, 2, \ldots, n-1$ we have

$$\begin{aligned} \sum_{i=1}^{n} u_i^{(h)} v_i &= \sum_{i=h+1}^{n} \frac{x^{i-h-1}}{(i-h-1)!} (-1)^{n-i} \frac{x^{n-i}}{(n-i)!}\\ &= \frac{x^{n-h-1}}{(n-h-1)!} \sum_{i=h+1}^{n} (-1)^{n-i} \binom{n-h-1}{i-h-1}\\ &= (-1)^{n-h-1} \frac{x^{n-h-1}}{(n-h-1)!} \sum_{k=0}^{n-h-1} (-1)^k \binom{n-h-1}{k}\\ &= \delta_{n-h-1,\,0} = \delta_{h,\,n-1}\,. \end{aligned}$$

In the case (2), having put

$$\omega(\alpha) = (\alpha - \alpha_1)(\alpha - \alpha_2) \cdots (\alpha - \alpha_n)\,, \tag{4}$$

we may assume

$$u_i(x) = e^{\alpha_i x}\,, \qquad v_i(x) = \frac{e^{-\alpha_i x}}{\omega'(\alpha_i)}\,, \qquad (i = 1, 2, \ldots, n)\,.$$

In fact there results

$$\sum_{i=1}^{n} u_i^{(h)} v_i = \sum_{i=1}^{n} \alpha_i^h e^{\alpha_i x} \frac{e^{-\alpha_i x}}{\omega'(\alpha_i)} = \sum_{i=1}^{n} \frac{\alpha_i^h}{\omega'(\alpha_i)}\,.$$

This expression is the sum of the residues of the rational function $\frac{\alpha^h}{\omega(\alpha)}$ at the poles of 1° order $\alpha_1, \alpha_2, \ldots, \alpha_n$; on the other hand the residue at the point $\alpha = \infty$ has value $-\delta_{h,n-1}$ and therefore the considered expression comes out to be equal to $\delta_{h,n-1}$.

Finally, in the case (3) we may assume, using position (4):

$$u_i(x) = x^{\alpha_i}, \qquad v_i(x) = \frac{x^{n-1-\alpha_i}}{\omega'(\alpha_i)}, \qquad (i = 1, 2, \ldots, n).$$

In fact we have

$$\sum_{i=1}^{n} u_i^{(h)} v_i = \sum_{i=1}^{n} \alpha_i (\alpha_i - 1) \cdots (\alpha_i - h + 1)\, x^{\alpha_i - h} \frac{x^{n-1-\alpha_i}}{\omega'(\alpha_i)}$$

$$= x^{n-h-1} \sum_{i=1}^{n} \frac{\alpha_i (\alpha_i - 1) \ldots (\alpha_i - h + 1)}{\omega'(\alpha_i)};$$

then it is sufficient to reason as before, considering the rational function $\frac{\alpha(\alpha - 1) \ldots (\alpha - h + 1)}{\omega(\alpha)}$.

2. Having called $W(x)$, $\overline{W}(x)$ the wronskian matrices of $U(x)$, $\overline{U}(x)$, we have $\overline{W}(x) = W(x)\, C^T$ and therefore $W^T(x) = C^{-1}\, \overline{W}^T(x)$. On the other hand $V(x)$ is determined by $V(x)\, W^T(x) = (0, \ldots, 0, 1)$ [see (1.3.5)] which can be written $[V(x)\, C^{-1}]\, \overline{W}^T(x) = (0, \ldots, 0, 1)$ and therefore $V(x)\, C^{-1}$ is the system associated to $\overline{U}(x)$, q.e.d.

3. By virtue of (1.1.7) the equation (2) is equivalent to

$$\sum_{i=0}^{k} (-1)^i \binom{n-i}{k-i} a_i^{(k-i)}(x) = a_k(x), \qquad (k = 0, 1, \ldots, n).$$

The formula with $k = 0$ is satisfied; the formulae with k even ($k = 2h$) and those ones with k odd ($k = 2h + 1$) can be respectively written:

$$\left.\begin{aligned} &\sum_{s=0}^{h-1} \binom{n-2s}{2h-2s} a_{2s}^{(2h-2s)}(x) - \sum_{r=0}^{h-1} \binom{n-2r-1}{2h-2r-1} a_{2r+1}^{(2h-2r-1)}(x) = 0, \\ &\qquad \left(h = 1, 2, \ldots, \left[\frac{n}{2}\right]\right), \end{aligned}\right\} \quad (4)$$

$$\left.\begin{aligned} &\sum_{s=0}^{h} \binom{n-2s}{2h-2s+1} a_{2s}^{(2h-2s+1)}(x) \\ &\quad - \sum_{r=0}^{h-1} \binom{n-2r-1}{2h-2r} a_{2r+1}^{(2h-2r)}(x) - 2\, a_{2h+1}(x) = 0, \\ &\qquad \left(h = 0, 1, \ldots, \left[\frac{n-1}{2}\right]\right). \end{aligned}\right\} \quad (5)$$

It is soon seen that from formula (5) we may obtain by recurrence $a_1(x)$, $a_3(x)$, $a_5(x)$, ... in terms of $a_0(x)$, $a_2(x)$, $a_4(x)$, ...

Therefore to prove formula (3) it is sufficient to verify that, substituting (3) in (4) and (5), these formulae reduce to identities in $a_0(x)$, $a_2(x)$, $a_4(x)$, ...

By means of such substitutions, formulae (4) and (5) become

$$\sum_{s=0}^{h-1}\binom{n-2s}{2h-2s} a_{2s}^{(2h-2s)}(x)\left[1-\sum_{r=s}^{h-1}\frac{2^{2r-2s+2}-1}{r-s+1}B_{2r-2s+2}\binom{2h-2s}{2r-2s+1}\right]=0,$$

$$\left(h=1,2,\ldots,\left[\frac{n}{2}\right]\right);$$

$$\sum_{s=0}^{h}\binom{n-2s}{2h-2s+1} a_{2s}^{(2h-2s+1)}(x)\left[1-\sum_{r=s}^{h-1}\frac{2^{2r-2s+2}-1}{r-s+1}B_{2r-2s+2}\binom{2h-2s+1}{2r-2s+1}\right.$$

$$\left.-2\,\frac{2^{2h-2s+2}-1}{h-s+1}B_{2h-2s+2}\right]=0,\quad \left(h=0,1,\ldots,\left[\frac{n-1}{2}\right]\right),$$

and that these are identities there depends on some properties of the Bernoulli's numbers which assure that the expression in brackets be all zeros (see § 3.10, Problem 1). See R. Suppa [1].

4. Having fixed a fundamental system $[u_1(x), \ldots, u_n(x)]$ of solutions of $E(u) = 0$, we soon see that the wanted condition is given by

$$\det\,[u_j^{(k)}(x_i)]_{\substack{i=1,\ldots,m;\ k=0,\ldots,p_i\\ j=1,\ldots,n}} \neq 0 \tag{3}$$

and that, calling $[\eta_{ik,j}]$ the inverse matrix of the one just considered, whence we have

$$\sum_{i=1}^{m}\sum_{k=0}^{p_i} u_l^{(k)}(x_i)\,\eta_{ik,j} = \delta_{lj}\,, \tag{4}$$

there results

$$u(x) = \sum_{i=1}^{m}\sum_{k=0}^{p_i} f^{(k)}(x_i)\,u_{ik}(x) \qquad \text{with} \qquad u_{ik}(x) = \sum_{j=1}^{n}\eta_{ik,j}\,u_j(x)\,. \tag{5}$$

In order to obtain the integral expression of the remainder $R(x)$, we denote by $[v_1(x), \ldots, v_n(x)]$ the fundamental system of solutions of the adjoint equation $E^*(v) = 0$, associated to $[u_1(x), \ldots, u_n(x)]$ and we introduce the following functions

$$v_{ik}(x) = \sum_{j=1}^{n} u_j^{(k)}(x_i)\,v_j(x)\,, \qquad (i = 1, \ldots, m;\quad k = 0, \ldots, p_i); \tag{6}$$

$$w_i(t, x) = \sum_{k=0}^{p_i} u_{ik}(x)\,v_{ik}(t)\,, \qquad (i = 1, \ldots, m)\,, \tag{7}$$

remarking that $w_i(t, x)$, *as function of* t, is solution of $E^*[v(t)] = 0$.

Moreover from (7), substituting in it the expressions of $u_{ik}(x)$ and $v_{ik}(t)$ given by (5) and (6), we get, using (4) and (1.3.18):

$$\sum_{i=1}^{m} w_i(t, x) = \sum_{j=1}^{n} u_j(x)\,v_j(t) = -\,K^*(t, x)\,. \tag{8}$$

Let us now write the Green-Lagrange identity (1.2.2), replacing x by t, v by $w_i(t, x)$, u by $f(t)$; we obtain

$$w_i(t, x)\, E[f(t)] = \frac{d}{dt} \sum_{h=0}^{n-1} f^{(h)}(t)\, E^*_{n-h-1}[w_i(t, x)]\,, \qquad (i = 1, \ldots, m)\,,$$

from which, integrating between x_i and x and then summing up with respect to i, we get

$$\sum_{i=1}^{m} \int_{x_i}^{x} w_i(t, x)\, E[f(t)]\, dt = \sum_{h=0}^{n-1} f^{(h)}(x) \left\{E^*_{n-h-1}\left[\sum_{i=1}^{m} w_i(t, x)\right]\right\}_{t=x}$$
$$- \sum_{i=1}^{m} \sum_{h=0}^{n-1} f^{(h)}(x_i)\, \{E^*_{n-h-1}[w_i(t, x)]\}_{t=x_i}\,. \tag{9}$$

For (8) and the well known properties of the resolvent kernel we have

$$\left\{E^*_{n-h-1}\left[\sum_{i=1}^{m} w_i(t, x)\right]\right\}_{t=x} = -\, \{E^*_{n-h-1}[K^*(t, x)]\}_{t=x}$$
$$= \begin{cases} -(-1)^{n-1} \cdot (-1)^n = 1\,, & (h = 0) \\ 0 & , \quad (h = 1, \ldots, n-1) \end{cases} \tag{10}$$

and, using successively (7), (6) and (1.3.9), we easy see that

$$\{E^*_{n-h-1}[w_i(t, x)]\}_{t=x_i} = \sum_{k=0}^{p_i} u_{ik}(x)\, \delta_{kh} = \begin{cases} u_{ih}(x)\,, & (h = 0, \ldots, p_i) \\ 0 & , \quad (h = p_i + 1, \ldots, n-1) \end{cases} \tag{11}$$

so that from (9), (10), (11) there follows

$$\sum_{i=1}^{m} \int_{x_i}^{x} w_i(t, x)\, E[f(t)]\, dt = f(x) - \sum_{i=1}^{m} \sum_{h=0}^{p_i} f^{(h)}(x_i)\, u_{ih}(x)\,.$$

From this formula and from (2), (5) we obtain

$$R(x) = \sum_{i=1}^{m} \int_{x_i}^{x} w_i(t, x)\, E[f(t)]\, dt\,.$$

The procedure here illustrated includes all the interpolation classical formulae; it is due to L. Gori [1].

6.2. Solutions to problem of § 2.7

1. This question is a Gauss problem (see § 2.5) with $g(x) = 1$, $m = 3$, $n = 4$, $E = \frac{d^4}{dx^4}$, $p = 3$. For theorem 2.5.I we must consider the boundary problem

$$\frac{d^4u}{dx^4} = 0\,, \qquad u(x_1) = u(x_2) = u(x_3) = 0$$

which has only one non trivial solution $U_1(x) = (x - x_1)(x - x_2)(x - x_3)$. Therefore, (1) is possible if and only if the nodes x_1, x_2, x_3 are such as to satisfy

$$\int_a^b (x - x_1)(x - x_2)(x - x_3)\, dx = 0\,. \tag{2}$$

Putting

$$x_i = \frac{a+b}{2} + \frac{b-a}{2} t_i \,, \qquad (i = 1, 2, 3; \quad -1 \leqq t_1 < t_2 < t_3 \leqq 1) \qquad (3)$$

and operating the substitution $x = \frac{a+b}{2} + \frac{b-a}{2} t$, (2) becomes $\int_{-1}^{1} (t - t_1) \times (t - t_2)(t - t_3)\, dt = 0$, that is $3\, t_1 t_2 t_3 + t_1 + t_2 + t_3 = 0$.

There are infinite ways of choosing t_1, t_2, t_3; for every choice we shall then obtain the nodes x_1, x_2, x_3 by means of (3) and in correspondence with them it will be possible to construct (1) in one and only one way because $q = 1$ and $m(n-p) - n + q = 3 \cdot 1 - 4 + 1 = 0$.

For instance, assuming $t_1 = -1$, $t_3 = 1$ we find $t_2 = 0$ and (1) becomes the Cavalieri-Simpson formula (see § 4.2).

2. It suffices to reason similarly as in § 2.5, with obvious variations, to establish theorem 2.5.I. Analogous considerations to those done for (2.5.8) hold for (3).

3. The problem here is a generalized Gauss problem of the type considered in the preceding problem, with $g(x) = 1$, $m = 2$, $n = 6$, $E = \frac{d^6}{dx^6}$, $p_1 = 3$, $p_2 = 5$. We have then to consider the boundary problem

$$\frac{d^6 u}{dx^6} = 0; \qquad u(x_1) = u'(x_1) = u''(x_1) = 0\,, \qquad u(x_2) = 0$$

which has the two non trivial solutions

$$U_r(x) = x^{r-1}(x - x_1)^3 (x - x_2)\,, \qquad (r = 1, 2)\,.$$

Therefore we must choose the nodes in a way that there results

$$\int_{-1}^{1} (x - x_1)^3 (x - x_2)\, dx = 0\,, \qquad \int_{-1}^{1} x(x - x_1)^3 (x - x_2)\, dx = 0$$

that is

$$\frac{1}{5} + x_1^2 + x_1 x_2 + x_1^3 x_2 = 0\,, \quad \frac{3}{5} x_1 + \frac{1}{5} x_2 + \frac{1}{3} x_1^3 + x_1^2 x_2 = 0\,.$$

We find the two solutions

$$\left(x_1 = -\frac{\sqrt{5}}{5}\,, \quad x_2 = \frac{\sqrt{5}}{3}\right), \quad \left(x_1 = \frac{\sqrt{5}}{5}\,, \quad x_2 = -\frac{\sqrt{5}}{3}\right)$$

to each of them there corresponds only one formula (1) since

$$m\,n - \sum p_i - n + q = 2 \cdot 6 - 8 - 6 + 2 = 0\,.$$

4. The construction of (1) yields a Gauss problem with $m = 3$, $n = 5$, $E = \frac{d^5}{dx^5}$, $p = 4$. We have therefore to consider the boundary problem

$$\frac{d^5 u}{dx^5} = 0; \qquad u(-1) = u(0) = u(1) = 0$$

which has the two non trivial solutions $U_1(x) = x(1 - x^2)$, $U_2(x) = x^2(1 - x^2)$, whence the weight $g(x)$ must satisfy the two conditions

$$\int_{-1}^{1} g(x)\, x\,(1 - x^2)\, dx = 0\,, \qquad \int_{-1}^{1} g(x)\, x^2\,(1 - x^2)\, dx = 0\,. \tag{2}$$

In order to solve (2) we put

$$g(x) = c_1\, x\,(1 - x^2) + c_2\, x^2\,(1 - x^2) + G(x) \tag{3}$$

where c_1, c_2 are constants and $G(x) \in L[-1, 1]$. Replacing (3) in (2) we obtain

$$c_1 = -\frac{105}{16} \int_{-1}^{1} \xi\,(1 - \xi^2)\, G(\xi)\, d\xi\,, \qquad c_2 = -\frac{315}{16} \int_{-1}^{1} \xi^2\,(1 - \xi^2)\, G(\xi)\, d\xi\,,$$

so that we conclude that

$$g(x) = G(x) - \frac{105}{16}\, x\,(1 - x^2) \int_{-1}^{1} \xi\,(1 - \xi^2)\, G(\xi)\, d\xi$$

$$- \frac{315}{16}\, x^2\,(1 - x^2) \int_{-1}^{1} \xi^2\,(1 - \xi^2)\, G(\xi)\, d\xi \tag{4}$$

where $G(x)$ is arbitrary.

For instance, with $G(x) = 1$ we obtain $g(x) = 1 - \frac{21}{4} x^2 (1 - x^2)$; with $G(x) = x$ we obtain $g(x) = \frac{1}{4} x\,(7\,x^2 - 3)$; and so on. In correspondence with each of these infinite possible choices of the weight $g(x)$ it is moreover possible to construct (1) in one and only one way since $m\,(n - p) - n + q = 3 \cdot 1 - 5 + 2 = 0$.

6.3. Solutions to problems of § 3.10

1. From (3.1.3), (3.1.4) there follows

$$\sum_{s=0}^{\infty} B_{2s+2} \frac{z^{2s+2}}{(2\,s + 2)!} = \frac{z}{e^z - 1} - 1 + \frac{1}{2} z\,, \qquad (|z| < 2\,\pi)\,,$$

and therefore writing $2\,z$ instead of z

$$\sum_{s=0}^{\infty} 2^{2s+2} B_{2s+2} \frac{z^{2s+2}}{(2\,s + 2)!} = \frac{2\,z}{e^{2z} - 1} - 1 + z\,, \qquad (|z| < \pi)\,.$$

Subtracting member by member we get

$$\sum_{s=0}^{\infty} (2^{2s+2} - 1)\, B_{2s+2} \frac{z^{2s+2}}{(2\,s + 2)!} = \frac{z}{2} \tanh \frac{z}{2}\,, \qquad (|z| < \pi);$$

which evidently is equivalent to (1).

Formula (1) can also be written

$$\cosh z - 1 = \sinh z \cdot \sum_{s=0}^{\infty} c_s \frac{z^{2s+1}}{(2\,s + 1)!}$$

and whence, substituting cosh z and sinh z by their developments in power series of z, we easy obtain (2).

Differentiating formula (1) we get

$$\frac{1}{\cosh z + 1} = \sum_{s=0}^{\infty} c_s \frac{z^{2s}}{(2s)!} \quad \text{or} \quad (1 + \cosh z) \sum_{s=0}^{\infty} c_s \frac{z^{2s}}{(2s)!} = 1$$

and whence, substituting cosh z by its development in series, we get

$$\sum_{s=0}^{p-1} c_s \binom{2p}{2s} + 2c_p = 0$$

which, summed up to (2), yields (3).

2. It is known that [see (3.1.3), (3.1.4)]:

$$\frac{z}{2} + \frac{z}{e^z - 1} = \sum_{n=0}^{\infty} B_{2n} \frac{z^{2n}}{(2n)!}, \quad (|z| < 2\pi).$$

Replacing $2iz$ in z we get (1); from (1) we pass to (2) considering that $\tan z = \text{cotan}\, z - 2 \,\text{cotan}\, 2z$.

3. It suffices to compare (3.1.1) with

$$\frac{z e^{xz}}{e^z - 1} = \frac{z}{e^z - 1} e^{xz} = \sum_{r=0}^{\infty} B_r \frac{z^r}{r!} \sum_{k=0}^{\infty} \frac{x^k z^k}{k!} = \sum_{n=0}^{\infty} \frac{z^n}{n!} \sum_{k=0}^{n} \frac{n!}{(n-k)!\,k!} B_{n-k} x^k.$$

4. From formula (3.1.1) changing x in mx we obtain

$$\frac{z e^{mxz}}{e^z - 1} = \sum_{n=0}^{\infty} B_n(mx) \frac{z^n}{n!}; \tag{2}$$

but it can be written

$$\frac{z e^{mxz}}{e^z - 1} = \frac{z e^{mxz}}{e^{mz} - 1} \frac{e^{mz} - 1}{e^z - 1} = \frac{z e^{mxz}}{e^{mz} - 1} \sum_{k=0}^{m-1} e^{kz} = \frac{1}{m} \sum_{k=0}^{m-1} \frac{mz e^{\left(x + \frac{k}{m}\right) mz}}{e^{mz} - 1}$$

and applying again (3.1.1), changing x in $x + \frac{k}{m}$ and z in mz:

$$\frac{z e^{mxz}}{e^z - 1} = \frac{1}{m} \sum_{k=0}^{m-1} \sum_{n=0}^{\infty} B_n\left(x + \frac{k}{m}\right) \frac{(mz)^n}{n!}. \tag{3}$$

From the comparison between (2) and (3) there follows (1).

5. For $0 < \alpha < 1$ we have $\Gamma(\alpha) = \int_0^{+\infty} x^{\alpha-1} e^{-x}\, dx > \int_0^1 x^{\alpha-1} e^{-x}\, dx$ $> e^{-1} \int_0^1 x^{\alpha-1}\, dx = \frac{1}{e\alpha}$.

6. Writing

$$\Gamma(\alpha) = \int_0^n x^{\alpha-1} e^{-x}\, dx + \int_n^{+\infty} x^{\alpha-1} e^{-x}\, dx$$

and considering that

$$\left(1-\frac{x}{n}\right)^n \leqq e^{-x} \leqq \left(1-\frac{x}{n}\right)^n + \frac{x^2 e^{-x}}{n}$$

it is first of all easy to prove that

$$\Gamma(\alpha) = \lim_{n\to\infty} \int_0^n x^{\alpha-1}\left(1-\frac{x}{n}\right)^n dx\,.$$

Then operating the substitution $x = n\,t$ and using (3.2.9), (3.2.10) we have:

$$\Gamma(\alpha) = \lim_{n\to\infty} n^\alpha \int_0^1 t^{\alpha-1}(1-t)^n\,dt = \lim_{n\to\infty} n^\alpha B(\alpha, n+1) = \lim_{n\to\infty} n^\alpha \frac{\Gamma(\alpha)\,\Gamma(n+1)}{\Gamma(\alpha+n+1)}$$

and this is, for (3.2.5), (3.2.6), equivalent to (1).

7. We may assume $\alpha > -k-1$ and therefore assume in (1) $n = k+1$; then

$$\lim_{\alpha\to -k} (\alpha+k)\,\Gamma(\alpha) = \lim_{\alpha\to -k} \frac{\Gamma(\alpha+k+1)}{\alpha(\alpha+1)\ldots(\alpha+k-1)}$$

$$= \frac{\Gamma(1)}{(-k)(-k+1)\ldots(-1)} = \frac{1}{(-1)^k\,k!}\,.$$

8. It suffices to apply the Hölder inequality

$$\left|\int_a^b f(t)\,g(t)\,dt\right| \leqq \left(\int_a^b |f(t)|^p\,dt\right)^{1/p}\left(\int_a^b |g(t)|^q\,dt\right)^{1/q}, \qquad \left(\frac{1}{p}+\frac{1}{q}=1\right),$$

putting $a = 0$, $b = +\infty$, $p = \dfrac{1}{\alpha}$, $q = \dfrac{1}{1-\alpha}$, $f(t) = e^{-\alpha t}\,t^{\alpha x}$, $g(t) = e^{-(1-\alpha)t} \times t^{(1-\alpha)(x-1)}$. In such a way we obtain

$$\int_0^{+\infty} e^{-t}\,t^{x+\alpha-1}\,dt \leqq \left(\int_0^{+\infty} e^{-t}\,t^x\,dt\right)^\alpha \left(\int_0^{+\infty} e^{-t}\,t^{x-1}\,dt\right)^{1-\alpha}$$

that is $\Gamma(x+\alpha) \leqq [\Gamma(x+1)]^\alpha\,[\Gamma(x)]^{1-\alpha}$ which, for $\Gamma(x+1) = x\,\Gamma(x)$, is equivalent to (1).

9. Formula (1) is an immediate consequence of (3.4.6), (3.4.8), (3.4.9).

10. Formulae (1) are consequences of the fact that

$$\lim_{\alpha\to+\infty} \frac{P_n^{(\alpha,\beta)}(x)}{\alpha^n} = \frac{1}{n!}\left(\frac{1+x}{2}\right)^n, \qquad \lim_{\beta\to+\infty} \frac{P_n^{(\alpha,\beta)}(x)}{\beta^n} = \frac{(-1)^n}{n!}\left(\frac{1-x}{2}\right)^n,$$

as there soon follows from (3.4.2) considering that we have

$$\lim_{\alpha\to+\infty} \frac{1}{\alpha^n}\binom{\alpha+n}{k} = \begin{cases} 0 & (k<n) \\ \dfrac{1}{n!} & (k=n) \end{cases};$$

$$\lim_{\beta\to+\infty} \frac{1}{\beta^n}\binom{\beta+n}{n-k} = \begin{cases} \dfrac{1}{n!} & (k=0) \\ 0 & (k>0) \end{cases}.$$

11. In the case $n = 1$, system (1) reduces to the only equation

$$\varphi_0(x_{11}) = \int_a^b p(x)\,(x - x_{11})^{2s+1}\,dx = 0$$

and since we have

$$\varphi_0(a) > 0\,, \quad \varphi_0(b) < 0\,, \quad \varphi_0'(x_{11}) = -(2s+1)\int_a^b p(x)\,(x - x_{11})^{2s}\,dx < 0\,,$$

it is clear that such an equation has one and only one solution x_{11} with $a < x_{11} < b$ [with any weight $p(x) \geqq 0$]. Let us now assume to have already proved that the following system [obtained from (1) by changing n in $n-1$]:

$$\int_a^b p(x) \cdot \prod_{i=1}^{k} (x - x_{n-1,i}) \cdot \prod_{j=1}^{n-1} (x - x_{n-1,j})^{2s+1}\,dx = 0\,, \quad (k = 0, 1, \ldots, n-2)\,, \tag{3}$$

has one and only one solution $x_{n-1,1}, x_{n-1,2}, \ldots, x_{n-1,n-1}$ with $a < x_{n-1,1} < \cdots < x_{n-1,n-1} < b$ [with any weight $p(x) \geqq 0$].

Let us consider system (1) and let us note that we may write it in the following way

$$\left.\begin{aligned}
&\varphi_0(x_{n1}, x_{n2}, \ldots, x_{nn}) = \int_a^b p(x) \prod_{j=1}^{n} (x - x_{nj})^{2s+1}\,dx = 0\,,\\
&\varphi_k(x_{n1}, x_{n2}, \ldots, x_{nn}) = \int_a^b p(x)\,(x - x_{n1})^{2s+2}\\
&\qquad \times \prod_{i=2}^{k} (x - x_{ni}) \cdot \prod_{j=2}^{n} (x - x_{nj})^{2s+1}\,dx = 0\,,\\
&\qquad (k = 1, 2, \ldots, n-1);
\end{aligned}\right\} \tag{4}$$

we then see that, for every fixed x_{n1} (with $a \leqq x_{n1} \leqq b$) its last $n-1$ equations form a system (in $n-1$ unknowns $x_{n2}, \ldots, x_{nn}$) of the type (3) and precisely obtained from (3) by changing $p(x)$ into $p(x)\,(x - x_{n1})^{2s+2}$, $x_{n-1,1}$ in $x_{n2}, \ldots, x_{n-1,n-1}$ in x_{nn}.

Under the hypothesis done for (3) we may then say that the last $n-1$ equations of the system (4) define a unique solution (depending on the choice of x_{n1}):

$$\left.\begin{aligned}
&x_{n2} = \xi_2(x_{n1})\,, \quad x_{n3} = \xi_3(x_{n1}), \ldots, x_{nn} = \xi_n(x_{n1})\,,\\
&\text{with} \quad a < \xi_2(x_{n1}) < \xi_3(x_{n1}) < \cdots < \xi_n(x_{n1}) < b\,.
\end{aligned}\right\} \tag{5}$$

Formulae (5) define in the Euclidean space R^n (with the coordinates $x_{n1}, x_{n2}, \ldots, x_{nn}$) a continuous arc of algebraic curve γ, from the point

$$\left.\begin{aligned}
&A[x_{n1} = a\,, \quad x_{n2} = \xi_2(a)\,, \quad x_{n3} = \xi_3(a), \ldots, x_{nn} = \xi_n(a)]\\
&\text{with} \quad a < \xi_2(a) < \xi_3(a) < \cdots < \xi_n(a) < b\,,
\end{aligned}\right\} \tag{6}$$

to the point

$$\left.\begin{aligned}
&B[x_{n1} = b\,, \quad x_{n2} = \xi_2(b)\,, \quad x_{n3} = \xi_3(b), \ldots, x_{nn} = \xi_n(b)]\\
&\text{with} \quad a < \xi_2(b) < \xi_3(b) < \cdots < \xi_n(b) < b\,.
\end{aligned}\right\} \tag{7}$$

The point A is in the hemispace $x_{n1} < x_{n2}$, the point B is in the hemispace $x_{n1} > x_{n2}$. There follows that γ meets the hyperplane $x_{n1} = x_{n2}$ (necessarily a finite number of times); let be

$$\left. \begin{array}{l} C\,[x_{n1} = c\,, \quad x_{n2} = c\,, \quad x_{n3} = \xi_3(c), \ldots, x_{nn} = \xi_n(c)] \\ \text{with} \quad a < c < \xi_3(c) < \cdots < \xi_n(c) < b\,, \end{array} \right\} \tag{8}$$

the first meet point (x_{n1} is increasing) of γ with such an hyperplane.

At all points of the arc $A\,C$ the last $n - 1$ equations of the system (4) are satisfied; *let us show that on such an arc there is at least one point at which there is also satisfied the first equation* $\varphi_0(x_{n1}, x_{n2}, \ldots, x_{nn}) = 0$.

We put

$$\prod_{j=2}^{n} (x - x_{nj}) = \sum_{k=0}^{n-1} \alpha_k(x_{n1}, x_{n2}, \ldots, x_{nn}) \prod_{i=1}^{k} (x - x_{ni})\,, \tag{9}$$

remarking that we have

$$\alpha_0(x_{n1}, x_{n2}, \ldots, x_{nn}) = \prod_{j=2}^{n} (x_{n1} - x_{nj})\,. \tag{10}$$

Since at the point A the equations $\varphi_1 = 0,\ \varphi_2 = 0, \ldots, \varphi_{n-1} = 0$ are satisfied it is clear that we may write

$$\alpha_0(A)\,\varphi_0(A) = \sum_{k=0}^{n-1} \alpha_k(A)\,\varphi_k(A)$$

that is, considering the expressions (1) of the functions φ_k and the values of the coordinates of A given by (6):

$$\begin{aligned} \alpha_0(A)\,\varphi_0(A) &= \sum_{k=0}^{n-1} \alpha_k[a, \xi_2(a), \ldots, \xi_n(a)] \\ &\quad \times \int_a^b p(x)\,(x - a) \prod_{i=2}^{k} [x - \xi_i(a)] \cdot (x - a)^{2s+1} \prod_{j=2}^{n} [x - \xi_j(a)]^{2s+1}\,dx\,. \end{aligned}$$

For (9) this formula can also be written

$$\begin{aligned} \alpha_0(A)\,\varphi_0(A) &= \int_a^b p(x) \prod_{j=2}^{n} [x - \xi_j(a)] \cdot (x - a)^{2s+1} \prod_{j=2}^{n} [x - \xi_j(a)]^{2s+1}\,dx \\ &= \int_a^b p(x)\,(x - a)^{2s+1} \prod_{j=2}^{n} [x - \xi_j(a)]^{2s+2}\,dx > 0 \end{aligned}$$

and there follows $\operatorname{sgn} \varphi_0(A) = \operatorname{sgn} \alpha_0[a, \xi_2(a), \ldots, \xi_n(a)]$ that is for (10) and (6):

$$\operatorname{sgn} \varphi_0(A) = \operatorname{sgn} \{[a - \xi_2(a)] \cdots [a - \xi_n(a)]\} = (-1)^{n-1}\,. \tag{11}$$

Let us now put

$$\prod_{j=3}^{n} (x - x_{nj}) = \sum_{k=0}^{n-2} \gamma(x_{n1}, x_{n2}, \ldots, x_{nn}) \prod_{i=1}^{k} (x - x_{ni})\,,$$

remarking that we have

$$\gamma_0(x_{n1}, x_{n2}, \ldots, x_{nn}) = \prod_{j=3}^{n} (x_{n1} - x_{nj})\,.$$

Then, operating as before, we may write

$$\gamma_0(C)\,\varphi_0(C) = \sum_{k=0}^{n-2} \gamma_k(C)\,\varphi_k(C) = \sum_{k=0}^{n-2} \gamma_k[c, c, \xi_3(c), \ldots, \xi_n(c)]$$

$$\times \int_a^b p(x) \cdot (x-c)^2 \prod_{i=3}^{k} [x - \xi_i(c)] \cdot (x-c)^{4s+2} \prod_{j=3}^{n} [x - \xi_j(c)]^{2s+1}\, dx$$

$$= \int_a^b p(x) \prod_{j=3}^{n} [x - \xi_j(c)] \cdot (x-c)^{4s+2} \prod_{j=3}^{n} [x - \xi_j(c)]^{2s+1}\, dx$$

$$= \int_a^b p(x)\,(x-c)^{4s+2} \prod_{j=3}^{n} [x - \xi_j(c)]^{2s+2}\, dx > 0$$

and therefore

$$\operatorname{sgn} \varphi_0(C) = \operatorname{sgn} \gamma_0(C) = \operatorname{sgn} \{[c - \xi_3(c)] \cdots [c - \xi_n(c)]\} = (-1)^{n-2}\,. \qquad (12)$$

From (11) and (12) there follows what we have affirmed, that is the existence of at least one solution of (1) satisfying (2). The uniqueness of the solution has been already proved in § 3.9. See A. Ossicini [1].

6.4. Solutions to problems of § 4.15

1. Assuming $\varphi_0(x) = -(x-a)$, $\varphi_1(x) = -(x-b)$, with the hypothesis $u(x) \in A\ C[a, b]$, we find again with (4.2.1)

$$R(u) = -\int_a^{(a+b)/2} (x-a)\,u'(x)\,dx - \int_{(a+b)/2}^{b} (x-b)\,u'(x)\,dx\,.$$

2. Assuming $\varphi_1(x) = -\left(x - \frac{a+b}{2}\right)$, with the hypothesis $u(x) \in A\ C[a, b]$, we find again (4.2.6) with

$$R(u) = \int_a^b \left(x - \frac{a+b}{2}\right) u'(x)\,dx\,.$$

3. If $E = \frac{d}{dx}$ assuming $\varphi_1(x) = -\left(x - \frac{5a+b}{6}\right)$, $\varphi_2(x) = -\left(x - \frac{a+5b}{6}\right)$, with the hypothesis $u(x) \in A\ C[a, b]$, we find again (4.2.11) with

$$R(u) = -\int_a^{(a+b)/2} \left(x - \frac{5a+b}{6}\right) u'(x)\,dx - \int_{(a+b)/2}^{b} \left(x - \frac{a+5b}{6}\right) u'(x)\,dx\,.$$

If $E = \frac{d^2}{dx^2}$, assuming $\varphi_1(x) = \frac{1}{2}(x-a)\left(x - \frac{2a+b}{3}\right)$, $\varphi_2(x) = \frac{1}{2}(x-b)$ $\times\left(x - \frac{a+2b}{3}\right)$, with the hypothesis $u(x) \in A\ C^1[a, b]$, we find again (4.2.11) with

$$R(u) = \frac{1}{2}\int_a^{(a+b)/2} (x-a)\left(x - \frac{2a+b}{3}\right) u''(x)\,dx$$

$$+ \frac{1}{2}\int_{(a+b)/2}^{b} (x-b)\left(x - \frac{a+2b}{3}\right) u''(x)\,dx\,.$$

If $E = \frac{d^3}{dx^3}$, assuming $\varphi_1(x) = -\frac{1}{6}(x-a)^2\left(x - \frac{a+b}{2}\right)$, $\varphi_2(x) = -\frac{1}{6}(x-b)^2 \times \left(x - \frac{a+b}{2}\right)$, with the hypothesis $u(x) \in A\, C^2[a, b]$, we find again (4.2.11) with

$$R(u) = -\frac{1}{6}\int_a^{(a+b)/2} (x-a)^2\left(x - \frac{a+b}{2}\right) u'''(x)\, dx$$
$$- \frac{1}{6}\int_{(a+b)/2}^{b} (x-b)^2\left(x - \frac{a+b}{2}\right) u'''(x)\, dx\,.$$

4. If in the Christoffel-Darboux formula (3.4.15) we put $y = x_{mj}^{(\alpha,\beta)}$ we obtain

$$\sum_{k=0}^{m-1} \frac{P_k^{(\alpha,\beta)}(x)\, P_k^{(\alpha,\beta)}(x_{mj}^{(\alpha,\beta)})}{h_k^{(\alpha,\beta)}} = -\frac{A_m^{(\alpha,\beta)}}{h_m^{(\alpha,\beta)}} \frac{P_m^{(\alpha,\beta)}(x)\, P_{m+1}^{(\alpha,\beta)}(x_{mj}^{(\alpha,\beta)})}{x - x_{mj}^{(\alpha,\beta)}}$$

wich for $x \to x_{mj}^{(\alpha,\beta)}$ becomes

$$\sum_{k=0}^{m-1} \frac{[P_k^{(\alpha,\beta)}(x_{mj}^{(\alpha,\beta)})]^2}{h_k^{(\alpha,\beta)}} = -\frac{A_m^{(\alpha,\beta)}}{h_m^{(\alpha,\beta)}} P_{m+1}^{(\alpha,\beta)}(x_{mj}^{(\alpha,\beta)}) \left[\frac{d}{dx} P_m^{(\alpha,\beta)}(x)\right]_{x = x_{mj}^{(\alpha,\beta)}}.$$

From it, there follows (1) since it is soon seen that, by virtue of (4.5.3) and (4.5.4), the second member has value $\frac{1}{H_{mj}^{(\alpha,\beta)}}$.

5. For $\alpha = \beta = \lambda - \frac{1}{2}$ formula (4.8.1) becomes

$$\int_{-1}^{1} (1-x^2)^{\lambda - \frac{1}{2}} u(x)\, dx = \left(\lambda + \frac{1}{2}\right) C_m^{(\lambda)} u(-1) + \sum_{i=1}^{m} C_{mi}^{(\lambda)} u(x_{mi}^{(\lambda+1)})$$
$$+ \left(\lambda + \frac{1}{2}\right) C_m^{(\lambda)} u(1) + R(u)\,, \tag{1}$$

where $x_{mi}^{(\lambda+1)} = x_{mi}^{\left(\lambda+\frac{1}{2},\, \lambda+\frac{1}{2}\right)}$ are the zeros of the Jacobi polynomial $P_m^{\left(\lambda+\frac{1}{2},\, \lambda+\frac{1}{2}\right)}(x)$ which coincide with the zeros of the ultraspherical polynomial $P_m^{(\lambda+1)}(x)$ [see (3.5.2)]; moreover we have put

$$C_m^{(\lambda)} = C_{m0}^{\left(\lambda-\frac{1}{2},\, \lambda-\frac{1}{2}\right)} = C_{m,\,m+1}^{\left(\lambda-\frac{1}{2},\, \lambda-\frac{1}{2}\right)} = \frac{2^{2\lambda} m! \left[\Gamma\left(\lambda + \frac{1}{2}\right)\right]^2}{\Gamma(2\lambda + m + 2)}\,,$$
$$\text{[see (4.8.5) and (4.8.6)]}, \tag{2}$$

$$C_{mi}^{(\lambda)} = C_{mi}^{\left(\lambda-\frac{1}{2},\, \lambda-\frac{1}{2}\right)} = \frac{2^{2\lambda}\left[\Gamma\left(\lambda + m + \frac{3}{2}\right)\right]^2}{(m+1)\,(m+1)!\,\Gamma(2\lambda + m + 2)\left[P_{m+1}^{\left(\lambda-\frac{1}{2},\, \lambda-\frac{1}{2}\right)}(x_{mi}^{(\lambda+1)})\right]^2}\,,$$
$$\text{[see (4.8.4)]}. \tag{3}$$

In formula (3) we can place, instead of $P_{m+1}^{\left(\lambda-\frac{1}{2},\lambda-\frac{1}{2}\right)}(x_{mi}^{(\lambda+1)})$, the ultraspherical polynomial $P_{m+1}^{(\lambda)}(x)$ according to the well known formula [see (3.5.2)]:

$$P_{m+1}^{\left(\lambda-\frac{1}{2},\lambda-\frac{1}{2}\right)}(x) = \frac{\Gamma(2\lambda)\,\Gamma\left(\lambda+m+\frac{3}{2}\right)}{\Gamma\left(\lambda+\frac{1}{2}\right)\Gamma(2\lambda+m+1)}\,P_{m+1}^{(\lambda)}(x);$$

using also (4.6.5) to express $\Gamma(2\lambda)$, we get

$$C_{mi}^{(\lambda)} = \frac{\pi}{2^{2\lambda-2}}\,\frac{\Gamma(2\lambda+m+1)}{(m+1)(m+1)!\,(2\lambda+m+1)\,[\Gamma(\lambda)]^2}\,\frac{1}{[P_{m+1}^{(\lambda)}(x_{mi}^{(\lambda+1)})]^2}\,. \tag{3'}$$

Formula (4.8.13) becomes

$$|R(u)| \leqq \frac{2^{2\lambda+2m+1}\,m!\left[\Gamma\left(\lambda+m+\frac{3}{2}\right)\right]^2\Gamma(2\lambda+m+2)}{(2m+2)!\,(\lambda+m+1)\,[\Gamma(2\lambda+2m+2)]^2}\,M_{2m+2} \tag{4}$$

and using again (4.6.5) to express $\Gamma(2\lambda+2m+2)$:

$$|R(u)| \leqq \frac{\pi}{2^{2\lambda+2m+1}}\,\frac{m!\,\Gamma(2\lambda+m+2)}{(2m+2)!\,(\lambda+m+1)\,[\Gamma(\lambda+m+1)]^2}\,M_{2m+2}\,. \tag{4'}$$

6. Formulae (2) follow soon from (1) remarking that $\prod_{i=1}^{m+1}(t-i+1) = (t-m)\prod_{i=1}^{m}(t-i+1)$ and therefore

$$\sum_{i=0}^{m} S_{i,m+1}\,t^{m+1-i} = (t-m)\sum_{i=0}^{m-1} S_{im}\,t^{m-i}$$

that is

$$\sum_{i=0}^{m} S_{i,m+1}\,t^{m+1-i} = \sum_{i=0}^{m} S_{im}\,t^{m+1-i} - m\sum_{i=1}^{m} S_{i-1,m}\,t^{m+1-i}\,.$$

We then note that from (4.11.2) and (4.11.7) it can be obtained

$$\omega_m'(x_{mj}) = \prod_{\substack{i=1\\(i\neq j)}}^{m}(x_{mj}-x_{mi}) = \prod_{\substack{i=1\\(i\neq j)}}^{m}[(j-i)\,h_m] = h_m^{m-1}(j-1)!\,(-1)^{m-j}\,(m-j)!\,.$$

Taking this into account and operating in (4.11.8) the substitution $x = -1 + h_m t$, we obtain

$$A_{mj} = \frac{(-1)^{m-j}}{h_m^{m-1}\,(m-j)!\,(j-1)!}\int_0^{m-1}\frac{\prod_{i=1}^{m}[(t-i+1)\,h_m]}{(t-j+1)\,h_m}\,h_m\,dt$$

$$= h_m\,\frac{(-1)^{m-j}}{(m-j)!\,(j-1)!}\int_0^{m-1}\frac{\sum_{i=0}^{m-1} S_{im}\,t^{m-i}}{t-j+1}\,dt$$

$$= \frac{2}{m-1}\,\frac{(-1)^{m-j}}{(m-j)!\,(j-1)!}\int_0^{m-1}\sum_{i=0}^{m-1}\left[\sum_{k=0}^{i} S_{km}\,(j-1)^{i-k}\right]t^{m-1-i}\,dt$$

and therefore

$$A_{mj} = 2\,\frac{(-1)^{m-j}}{(m-j)!\,(j-1)!}\sum_{i=0}^{m-1}\frac{(m-1)^{m-1-i}}{(m-i)!}\sum_{k=0}^{i} S_{km}(j-1)^{i-k}\,.$$

7. We observe first of all that, putting $x = -1 + h_m t$, we obtain immediately

$$b_m = \frac{h_m^{m+1}}{m!}\int_0^{m-1}\prod_{i=1}^{m}(t-i+1)\,dt\,, \qquad (m \text{ even}). \tag{3}$$

In the case that m is odd, considering that $\omega_m(x)$ is an odd function, we may write of all

$$b_m = \frac{1}{(m+1)!}\int_{-1}^{1}(x+1-m\,h_m)\,\omega_m(x)\,dx\,, \qquad (m \text{ odd}),$$

and whence, operating as before, deduce

$$b_m = \frac{h_m^{m+2}}{(m+1)!}\int_0^{m-1}\prod_{i=1}^{m+1}(t-i+1)\,dt\,, \qquad (m \text{ odd}). \tag{4}$$

Then, *if m is even*, we may write

$$(m+1)\int_0^{m-1}\prod_{i=1}^{m}(t-i+1)\,dt = \int_0^{m-1}[(t+1)-(t-m)]\prod_{i=1}^{m}(t-i+1)\,dt$$

$$= \int_0^{m-1}\left[\prod_{i=0}^{m}(t-i+1) - \prod_{i=1}^{m+1}(t-i+1)\right]dt$$

$$= \int_0^{m-1}\left[\prod_{i=1}^{m+1}(t-i+2) - \prod_{i=1}^{m+1}(t-i+1)\right]dt$$

$$= \sum_{k=0}^{m-2}\int_k^{k+1}\left[\prod_{i=1}^{m+1}(t-i+2) - \prod_{i=1}^{m+1}(t-i+1)\right]dt$$

$$= \sum_{k=0}^{m-2}\int_0^{1}\left[\prod_{i=1}^{m+1}(\tau-i+k+2) - \prod_{i=1}^{m+1}(\tau-i+k+1)\right]d\tau$$

$$= \int_0^{1}\left[\sum_{k=1}^{m-1}\prod_{i=1}^{m+1}(\tau-i+k+1) - \sum_{k=0}^{m-2}\prod_{i=1}^{m+1}(\tau-i+k+1)\right]d\tau$$

$$= \int_0^{1}\left[\prod_{i=1}^{m+1}(\tau-i+m) - \prod_{i=1}^{m+1}(\tau-i+1)\right]d\tau$$

$$= \int_0^{1}\prod_{i=1}^{m+1}(1-\tau-i+m)\,d\tau - \int_0^{1}\prod_{i=1}^{m+1}(\tau-i+1)\,d\tau\,.$$

If in the first integral we change the sign of every factor of the shown product, such a product changes its sign (being $m+1$ odd) and therefore we have

$$(m+1)\int_0^{m-1}\prod_{i=1}^{m}(t-i+1)\,dt$$
$$= -\int_0^1\prod_{i=1}^{m+1}[\tau-(m-i+2)+1]\,d\tau - \int_0^1\prod_{i=1}^{m+1}(\tau-i+1)\,d\tau;$$

but the last two products are evidently equal and therefore, recalling (1), we conclude that

$$(m+1)\int_0^{m-1}\prod_{i=1}^{m}(t-i+1)\,dt = -2\,(m+1)!\,c_{m+1}\,, \quad (m \text{ even}). \tag{5}$$

From (3) and (5) there follows surely (2) in the case of m even. Let us now assume m *odd* and let us remark that

$$\int_0^{m-1}\prod_{i=1}^{m+1}(t-i+1)\,dt = \int_0^{m}\prod_{i=1}^{m+1}(t-i+1)\,dt - \int_{m-1}^{m}\prod_{i=1}^{m+1}(t-i+1)\,dt\,. \tag{6}$$

We may apply, to the first integral of the second member, formula (5) with $m+1$ (even) instead of m and therefore it has value $-2\,(m+1)!\,c_{m+2}$. The second integral becomes as follows

$$\int_0^1\prod_{i=1}^{m+1}(\tau-i+m)\,d\tau = \int_0^1\prod_{i=1}^{m+1}(1-\tau-i+m)\,d\tau$$

and successively, changing the sign of every factor of the last product (which does not change it, being $m+1$ even):

$$\int_0^1\prod_{i=1}^{m+1}[\tau-(m-i+2)+1]\,d\tau = \int_0^1\prod_{j=1}^{m+1}(\tau-j+1)\,d\tau = (m+1)!\,c_{m+1}\,.$$

Therefore (6) becomes

$$\int_0^{m-1}\prod_{i=1}^{m+1}(t-i+1)\,dt = -2\,(m+1)!\,c_{m+2} - (m+1)!\,c_{m+1};$$

from it and from (4) there follows (2) in the case of m odd.

8. Let us begin by proving that, with any m, we have

$$|c_{m+1}| = \left|\frac{1}{(m+1)!}\int_0^1\prod_{i=1}^{m+1}(t-i+1)\,dt\right| < \frac{1}{m\,(\log m)^2}\,. \tag{2}$$

In fact, considering that

$$\prod_{i=1}^{m+1}(t-i+1) = t\,(t-1)\,(t-2)\cdots(t-m)$$
$$= (-1)^m\,t\cdot(m-t)\,(m-1-t)\cdots(1-t) = (-1)^m\,t\,\frac{\Gamma(m+1-t)}{\Gamma(1-t)}\,,$$

we deduce

$$|c_{m+1}| = \frac{1}{(m+1)!}\int_0^1\frac{t\,\Gamma(m+1-t)}{\Gamma(1-t)}\,dt < \frac{1}{(m+1)!}\int_0^1 t\,\Gamma(m+1-t)\,dt;$$

but we have $\Gamma(m+1-t) \leqq m^{1-t}\,\Gamma(m) = m^{-t}\,m!$ and therefore

$$|c_{m+1}| < \frac{1}{m+1}\int_0^1 t\,m^{-t}\,dt = \frac{1}{m+1}\left[\frac{1}{(\log m)^2} - \frac{1}{m\log m}\left(1+\frac{1}{\log m}\right)\right] < \frac{1}{m\,(\log m)^2}.$$

Having thus proved (2), from it and from (2) of the preceding problem, there follows (1) in the case of m even.

Then we have

$$\begin{aligned}|2\,c_{m+2}+c_{m+1}| &= \frac{1}{(m+2)!}\left|2\int_0^1 \prod_{i=1}^{m+2}(t-i+1)\,dt\right.\\ &\quad \left. + (m+2)\int_0^1 \prod_{i=1}^{m+1}(t-i+1)\,dt\right|\\ &= \frac{1}{(m+2)!}\left|2\int_0^1 (-1)^{m+1}\,t\,\frac{\Gamma(m+2-t)}{\Gamma(1-t)}\,dt\right.\\ &\quad \left. + (m+2)\int_0^1 (-1)^m\,t\,\frac{\Gamma(m+1-t)}{\Gamma(1-t)}\,dt\right|\\ &= \frac{1}{(m+2)!}\int_0^1 t\,\frac{\Gamma(m+1-t)}{\Gamma(1-t)}\,[2\,(m+1-t)-(m+2)]\,dt\\ &= \frac{1}{(m+2)!}\int_0^1 t\,\frac{\Gamma(m+1-t)}{\Gamma(1-t)}\,(m-2\,t)\,dt\\ &< \frac{m}{(m+2)!}\int_0^1 t\,\Gamma(m+1-t)\,dt\,,\end{aligned}$$

but we have seen before that

$$\frac{1}{(m+1)!}\int_0^1 t\,\Gamma(m+1-t)\,dt < \frac{1}{m\,(\log m)^2}$$

and therefore

$$|2\,c_{m+2}+c_{m+1}| < \frac{m}{(m+2)!}\,\frac{(m+1)!}{m\,(\log m)^2} < \frac{1}{m\,(\log m)^2}.$$

From it and from (2) of the preceding problem there follows (1) in the case of m odd.

9. We find

$$b_2 = \frac{1}{2!}\int_{-1}^{1}(x^2-1)\,dx = -\frac{2}{3}, \qquad b_3 = \frac{1}{4!}\int_{-1}^{1}(x^4-x^2)\,dx = -\frac{1}{90},$$

$$b_4 = \frac{1}{4!}\int_{-1}^{1}\left(x^4-\frac{10}{9}x^2+\frac{1}{9}\right)dx = -\frac{2}{405},$$

$$b_5 = \frac{1}{6!}\int_{-1}^{1}\left(x^6-\frac{5}{4}x^4+\frac{1}{4}x^2\right)dx = -\frac{1}{15120},$$

$$b_6 = \frac{1}{6!}\int_{-1}^{1}\left(x^6-\frac{7}{5}x^4+\frac{259}{625}x^2-\frac{9}{625}\right)dx = -\frac{22}{590625},$$

$$b_7 = \frac{1}{8!}\int_{-1}^{1}\left(x^8-\frac{14}{9}x^6+\frac{49}{81}x^4-\frac{4}{81}x^2\right)dx = -\frac{1}{3061800}.$$

10. Having fixed p (with $0 \leqq p \leqq 2m-2$), we associate to $u(x)$, which occurs in (4.6.25), the following function

$$\bar{u}_p(x) = \int_{-1}^{x}\frac{(x-\xi)^p}{p!}u^{(p+1)}(\xi)\,d\xi,$$

for which we have $\bar{u}_p^{(p+1)}(x) = u^{(p+1)}(x)$ (with $p+1 \leqq 2m-1$) and therefore $\bar{u}_p^{(2m)}(x) = u^{(2m)}(x)$.
For (2.4.8) we have $R(u) = R(\bar{u}_p)$, that is

$$R(u) = \int_{-1}^{1}\frac{dx}{\sqrt{1-x^2}}\int_{-1}^{x}\frac{(x-\xi)^p}{p!}u^{(p+1)}(\xi)\,d\xi - \frac{\pi}{m}\sum_{i=1}^{m}\int_{-1}^{x_i}\frac{(x_i-\xi)^p}{p!}u^{(p+1)}(\xi)\,d\xi. \tag{2}$$

On the other hand if we put in (4.6.25) $u(x) = \int_{-1}^{x}\frac{(x-\xi)^p}{p!}d\xi = \frac{(x+1)^{p+1}}{(p+1)!}$

(polynomial of degree $\leqq 2m-1$), we get

$$\int_{-1}^{1}\frac{dx}{\sqrt{1-x^2}}\int_{-1}^{x}\frac{(x-\xi)^p}{p!}d\xi = \frac{\pi}{m}\sum_{i=1}^{m}\int_{-1}^{x_i}\frac{(x_i-\xi)^p}{p!}d\xi.$$

The common value of these two expressions, which we shall denote by C_p, can be soon computed noting that $C_p = \int_{-1}^{1}\frac{(x+1)^{p+1}}{(p+1)!}\frac{dx}{\sqrt{1-x^2}}$ and therefore, operating the substitution $x = \cos 2t$:

$$C_p = \frac{2^{p+2}}{(p+1)!}\int_{0}^{\pi/2}\cos^{2p+2}t\,dt = \frac{2^{p+2}}{(p+1)!}\frac{(2p+1)!!}{(2p+2)!!}\frac{\pi}{2} = \pi\frac{(2p+1)!!}{[(p+1)!]^2}. \tag{3}$$

We examine again (2) and, putting $\lambda_{p+1} = \min\limits_{-1 \leqq x \leqq 1} u^{(p+1)}(x)$, $\Lambda_{p+1} = \max\limits_{-1 \leqq x \leqq 1} u^{(p+1)}(x)$, we remark that from it we deduce

$$\lambda_{p+1}\, C_p - \Lambda_{p+1}\, C_p \leqq R(u) \leqq \Lambda_{p+1}\, C_p - \lambda_{p+1}\, C_p$$

that is $|R(u)| \leqq C_p\, (\Lambda_{p+1} - \lambda_{p+1})$. Introducing the oscillation $\omega_{p+1} = \Lambda_{p+1} - \lambda_{p+1}$, we then have $|R(u)| \leqq C_p\, \omega_{p+1}$ and this formula, for (3), coincides with (1).

11. Formula (4.13.4) with $s = 0$ is written

$$\int\limits_a^b p(x)\, u(x)\, dx = \sum_{i=1}^{m} A_{0i}\, u(x_i) + R(u); \qquad [u^{(2m)}(x) = 0 \Rightarrow R(u) = 0]\,, \quad (1)$$

where $x_1, x_2, \ldots, x_m$ are the zeros of the polynomial $P_m(x)$ of the orthogonal system in $[a, b]$, connected to the weight $p(x)$. Putting in (1) $u(x) = \left[\frac{P_m(x)}{x - x_i}\right]^2$ (polynomial of degree $2\,m - 2$), we get

$$\int\limits_a^b p(x) \left[\frac{P_m(x)}{x - x_i}\right]^2 dx = A_{0i}[P'_m(x_i)]^2$$

from which we surely obtain $A_{0i} > 0$.

12. Having fixed ν, let us consider the polynomial $Q_\nu(x)$, of degree $(2\,s + 2)\,m - 2$, which is defined by the $(2\,s + 2)\,m - 1$ following conditions

$$Q_\nu(x_k) = 1\,, \quad (k = 1, 2, \ldots, \nu); \quad Q_\nu(x_k) = 0\,, \quad (k = \nu + 1,\ \nu + 2, \ldots, m)\,, \quad (2)$$

$$Q_\nu^{(h)}(x_k) = 0\,, \qquad (h = 1, 2, \ldots, 2\,s; \quad k = 1, 2, \ldots, m)\,, \quad (3)$$

$$Q_\nu^{(2s+1)}(x_k) = 0\,, \qquad (k = 1, \ldots, \nu - 1\,, \quad \nu + 1, \ldots, m)\,. \quad (4)$$

The polynomial $Q'_\nu(x)$ of degree $(2\,s + 2)\,m - 3$ has, for (3) and (4), $m - 1$ zeros, each one of multiplicity $2\,s + 1$, at the points $x_1, \ldots, x_{\nu-1}, x_{\nu+1}, \ldots, x_m$ and a zero of multiplicity $2\,s$ at the point x_ν; moreover, for (2) and for the Rolle theorem, it has at least one zero inside each of the $m - 2$ intervals, $[x_1, x_2], \ldots, [x_{\nu-1}, x_\nu], [x_{\nu+1}, x_{\nu+2}], \ldots, [x_{m-1}, x_m]$. These last $m - 2$

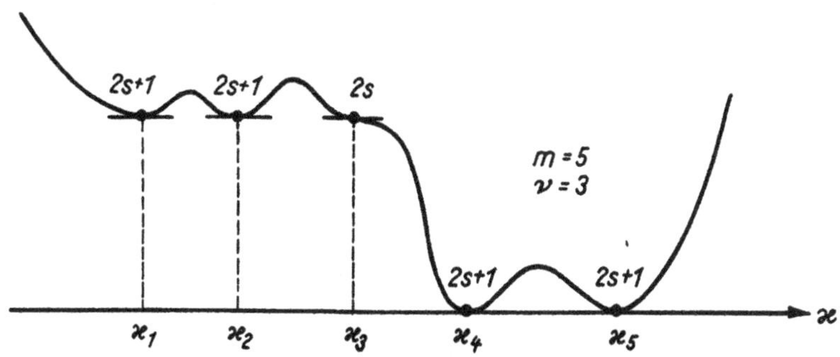

Fig. 6

zeros are necessarily simple, since in the whole we have already found for $Q_\nu'(x)$ a number of zeros equal to $(m-1)(2s+1)+2s+(m-2) = (2s+2)m-3$. There immediately follows that $Q_\nu(x)$ has a graph of the type of Fig. 6 and therefore we have

$$\left.\begin{array}{lll} Q_\nu(x) \geqq 1 & \text{for} & a \leqq x \leqq x_\nu\,, \\ Q_\nu(x) \geqq 0 & \text{for} & x_\nu \leqq x \leqq b\,.^{1)} \end{array}\right\} \tag{5}$$

If we put in (4.13.4) $u(x) = Q_\nu(x)$, we obtain

$$\int_a^b p(x)\, Q_\nu(x)\, dx = \sum_{i=1}^{\nu} A_{0i}$$

and since for (5) the integral in the first member is greater than $\int_a^{x_\nu} p(x)\, dx$ we *get the inequality which in* (1) *is written at left.*

Let us now consider the polynomial $\overline{Q}_\nu(x)$ of degree $(2s+2)m-2$ determined by the $(2s+2)m-1$ conditions

$$\overline{Q}_\nu(x_k) = 0\,, \quad (k = 1, 2, \ldots, \nu); \quad \overline{Q}_\nu(x_k) = 1\,, \quad (k = \nu+1\,, \nu+2, \ldots, m)\,,$$

$$\overline{Q}_\nu^{(h)}(x_k) = 0\,, \quad (h = 1, 2, \ldots, 2s; \quad k = 1, 2, \ldots, m)\,,$$

$$\overline{Q}_\nu^{(2s+1)}(x_k) = 0\,, \quad (k = 1, \ldots, \nu\,, \quad \nu+2, \ldots, m)\,.$$

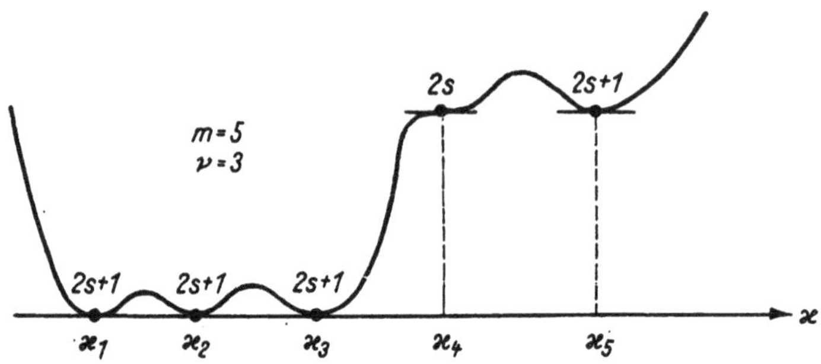

Fig. 7

Reasoning similarly as before we see that $\overline{Q}_\nu(x)$ has a graph like in Fig. 7, whence we have

$$\left.\begin{array}{lll} \overline{Q}_\nu(x) \geqq 0 & \text{for} & a \leqq x \leqq x_{\nu+1}\,, \\ \overline{Q}_\nu(x) \geqq 1 & \text{for} & x_{\nu+1} \leqq x \leqq b\,. \end{array}\right\} \tag{6}$$

Putting $u(x) = \overline{Q}_\nu(x)$ in (4.13.4), we get

$$\int_a^b p(x)\, \overline{Q}_\nu(x)\, dx = \sum_{i=\nu+1}^{m} A_{0i};$$

[1]) In Fig. 6 and 7, we have marked, besides those points of the graph where the tangent is parallel to the axis x, the order of the contact of the graph with the tangent.

but for (6) the integral in the first member is greater than $\int_{x_\nu+1}^{b} p(x)\,dx$ and therefore

$$\int_{x_\nu+1}^{b} p(x)\,dx < \sum_{i=\nu+1}^{m} A_{0i}\,. \tag{7}$$

On the other hand, if in (4.13.4) we put $u(x) = 1$, we may write $\int_a^b p(x)\,dx = \sum_{i=1}^{m} A_{0i}$ and if we subtract (7) from it we *obtain the inequality which in* (1) *is written at right.*

13. By virtue of theorem 2.5.I we must consider the boundary problem

$$\frac{d^{m(r+2)+2r+2}u}{dx^{m(r+2)+2r+2}} = 0;$$

$$u^{(h)}(a) = u^{(h)}(x_1) = \cdots = u^{(h)}(x_m) = u^{(h)}(b) = 0\,, \qquad (h = 0, 1, \ldots, r)$$

and its non trivial solutions which are

$$x^k \cdot (x-a)^{r+1} \cdot \prod_{i=1}^{m} (x-x_i)^{r+1} \cdot (b-x)^{r+1}\,, \qquad (k = 0, 1, \ldots, m-1)\,.$$

Therefore, (1) is possible if and only if there are satisfied

$$\left.\begin{array}{c} \int_a^b p(x)\,(x-a)^{r+1}\,(b-x)^{r+1} \cdot x^k \left[\prod_{i=1}^{m} (x-x_i)\right]^{r+1} dx = 0\,, \\ (k = 0, 1, \ldots, m-1)\,. \end{array}\right\} \tag{3}$$

We see that r must be even; having put $r = 2s$, formula (3) is written

$$\int_a^b p(x)\,(x-a)^{2s+1}\,(b-x)^{2s+1} \cdot x^k \left[\prod_{i=1}^{m} (x-x_i)\right]^{2s+1} dx = 0\,,$$
$$(k = 0, 1, \ldots, m-1)\,,$$

and this shows that *the nodes x_i must coincide with the zeros of the polynomial $P_{s,m}(x)$ of the s-orthogonal system relative to the weight $p(x)\,(x-a)^{2s+1}$ $(b-x)^{2s+1}$.*

With such a choice of the nodes the formula is unique since, with the notations of theorem 2.5.I we have

$$m\,(n-p) - n + q = (m+2)\,(2s+1) - [m\,(2s+2) + 4s + 2] - m = 0\,.$$

After that we may go on as in § 4.13 (see A. Ossicini [3]).

6.5. Solutions to problems of § 5.4

1. It is sufficient to operate as in § 5.1, assuming $\alpha_j = a + \frac{j}{\nu}(b-a)$ and therefore $\delta_j = \frac{1}{\nu}(b-a)$.

Since, instead of (5.1.3) we now have

$$\int_a^b u(x)\,dx = (b-a)\,u\left(\frac{a+b}{2}\right) + R(u) \tag{1}$$

with

$$R(u) = \frac{1}{2}\int_a^{(a+b)/2} (x-a)^2\,u''(x)\,dx + \frac{1}{2}\int_{(a+b)/2}^b (x-b)^2\,u''(x)\,dx\,, \tag{2}$$

formulae (5.1.4) and (5.1.5) are respectively written

$$\int_a^b u(x)\,dx = \frac{b-a}{\nu}\sum_{j=1}^{\nu} u\left[a + \frac{2j-1}{2\nu}(b-a)\right] + \varrho(u)\,, \tag{3}$$

$$\varrho(u) = \frac{1}{2\nu^3}\sum_{j=1}^{\nu}\left\{\int_a^{(a+b)/2} (x-a)^2\,u''\left[a + \frac{j-1}{\nu}(b-a) + \frac{1}{\nu}(x-a)\right]dx\right.$$

$$\left. + \int_{(a+b)/2}^b (x-b)^2\,u''\left[a + \frac{j-1}{\nu}(b-a) + \frac{1}{\nu}(x-a)\right]dx\right\}. \tag{4}$$

With procedure for $\varrho(u)$ equal to that done for $\varrho_0(u)$ in the proof of theor. 5.1.I, we find

$$|\varrho(u)| \leqq \frac{(b-a)^2}{8\nu^2}\int_a^b |u''(\xi)|\,d\xi\,; \tag{5}$$

assuming $M_2 = \sup_{a \leqq x \leqq b} |u''(x)| < +\infty$, we obtain from (4)

$$|\varrho(u)| \leqq \frac{(b-a)^3}{24\nu^2} M_2\,. \tag{6}$$

2. Operating as in the preceding problem and starting from

$$\int_a^b u(x)\,dx = \frac{b-a}{2}[u(a) + u(b)] + R(u)\,, \tag{1}$$

$$R(u) = \frac{1}{2}\int_a^b (x-a)(x-b)\,u''(x)\,dx\,, \tag{2}$$

we find successively

$$\int_a^b u(x)\,dx = \frac{b-a}{2\nu}\sum_{j=1}^{\nu}\left\{u\left[a + \frac{j-1}{\nu}(b-a)\right] + u\left[a + \frac{j}{\nu}(b-a)\right]\right\} + \varrho(u)$$

$$= \frac{b-a}{\nu}\left\{\frac{1}{2}u(a) + \sum_{j=1}^{\nu-1} u\left[a + \frac{j}{\nu}(b-a)\right] + \frac{1}{2}u(b)\right\} + \varrho(u)\,, \tag{3}$$

$$\varrho(u) = \frac{1}{2\nu^3} \sum_{j=1}^{\nu} \int_a^b (x-a)(x-b)\, u''\left[a + \frac{j-1}{\nu}(b-a) + \frac{1}{\nu}(x-a)\right] dx\,, \quad (4)$$

$$|\varrho(u)| \leqq \frac{(b-a)^2}{8\nu^2} \int_a^b |u''(\xi)|\, d\xi\,, \quad (5)$$

$$|\varrho(u)| \leqq \frac{(b-a)^3}{12\nu^2} M_2\,. \quad (6)$$

3. Operating as in the preceding two problems and starting from

$$\int_a^b u(x)\, dx = \frac{b-a}{6}\left[u(a) + 4u\left(\frac{a+b}{2}\right) + u(b)\right] + R(u)\,, \quad (1)$$

$$R(u) = \frac{1}{24} \int_a^{(a+b)/2} (x-a)^3 \left(x - \frac{a+2b}{3}\right) u''''(x)\, dx$$

$$+ \frac{1}{24} \int_{(a+b)/2}^{b} (x-b)^3 \left(x - \frac{2a+b}{3}\right) u''''(x)\, dx\,, \quad (2)$$

we find successively

$$\int_a^b u(x)\, dx = \frac{b-a}{6\nu} \sum_{j=1}^{\nu} \left\{u\left[a + \frac{j-1}{\nu}(b-a)\right] + 4u\left[a + \frac{2j-1}{2\nu}(b-a)\right]\right.$$

$$\left. + u\left[a + \frac{j}{\nu}(b-a)\right]\right\} + \varrho(u) = \frac{b-a}{6\nu}\left\{u(a) + u(b)\right.$$

$$\left. + 2\sum_{j=1}^{\nu-1} u\left[a + \frac{j}{\nu}(b-a)\right] + 4\sum_{j=1}^{\nu} u\left[a + \frac{2j-1}{2\nu}(b-a)\right]\right\} + \varrho(u)\,, \quad (3)$$

$$\varrho(u) = \frac{1}{24\nu^5} \sum_{j=1}^{\nu} \left\{\int_a^{(a+b)/2} (x-a)^3 \left(x - \frac{a+2b}{3}\right)\right.$$

$$\times u''''\left[a + \frac{j-1}{\nu}(b-a) + \frac{1}{\nu}(x-a)\right] dx + \int_{(a+b)/2}^{b} (x-b)^3 \times$$

$$\left. \times \left(x - \frac{2a+b}{3}\right) u''''\left[a + \frac{j-1}{\nu}(b-a) + \frac{1}{\nu}(x-a)\right] dx\right\}, \quad (4)$$

$$|\varrho(u)| \leqq \frac{(b-a)^4}{1152\nu^4} \int_a^b |u''''(\xi)|\, d\xi\,, \quad (5)$$

$$|\varrho(u)| \leqq \frac{(b-a)^5}{2880\,\nu^4} M_4\,, \qquad (M_4 = \sup_{a \leqq x \leqq b} |u''''(x)|)\,. \tag{6}$$

4. Operating as in the preceding three problems and starting from

$$\int_0^1 u(x)\,dx = \frac{1}{2}\,[u(0) + u(1)] + \sum_{k=1}^{n} \frac{B_{2k}}{(2\,k)!}\,[u^{(2k-1)}(0) - u^{(2k-1)}(1)] + R(u)\,, \tag{1}$$

$$R(u) = \int_0^1 \frac{B_{2n+2}(x) - B_{2n+2}}{(2\,n+2)!}\,u^{(2n+2)}(x)\,dx\,, \tag{2}$$

we find successively

$$\int_0^1 u(x)\,dx = \sum_{j=1}^{\nu} \left\{\frac{1}{2\,\nu}\left[u\left(\frac{j-1}{\nu}\right) + u\left(\frac{j}{\nu}\right)\right] + \sum_{k=1}^{n} \frac{1}{\nu^{2k}}\,\frac{B_{2k}}{(2\,k)!}\left[u^{(2k-1)}\left(\frac{j-1}{\nu}\right)\right.\right.$$
$$\left.\left. - u^{(2k-1)}\left(\frac{j}{\nu}\right)\right]\right\} + \varrho(u) = \frac{1}{\nu}\left[\frac{1}{2}\,u(0) + \sum_{j=1}^{\nu-1} u\left(\frac{j}{\nu}\right) + \frac{1}{2}\,u(1)\right]$$
$$+ \sum_{k=1}^{n} \frac{1}{\nu^{2k}}\,\frac{B_{2k}}{(2\,k)!}\,[u^{(2k-1)}(0) - u^{(2k-1)}(1)] + \varrho(u)\,, \tag{3}$$

$$\varrho(u) = \frac{1}{\nu^{2n+3}} \sum_{j=1}^{\nu} \int_0^1 \frac{B_{2n+2}(x) - B_{2n+2}}{(2\,n+2)!}\,u^{(2n+2)}\left(\frac{j-1}{\nu} + \frac{x}{\nu}\right) dx\,, \tag{4}$$

$$|\varrho(u)| \leqq \frac{1}{\nu^{2n+2}}\,2\left(1 - \frac{1}{2^{2n+2}}\right)\frac{|B_{2n+2}|}{(2\,n+2)!} \int_0^1 |u^{(2n+2)}(\xi)|\,d\xi\,, \tag{5}$$

$$|\varrho(u)| \leqq \frac{1}{\nu^{2n+2}}\,\frac{|B_{2n+2}|}{(2\,n+2)!}\,M_{2n+2}\,, \qquad \left(M_{2n+2} = \sup_{0 \leqq x \leqq 1} |u^{(2n+2)}(x)|\right). \tag{6}$$

5. Starting from the elementary formula (4.4.19), that is

$$\int_a^b \frac{u(x)\,dx}{\sqrt{x-a}} = \frac{2}{3}\sqrt{b-a}\;[2\,u(a) + u(b)] + R(u)\,; \tag{1}$$

with

$$R(u) = -\frac{4}{3} \int_a^b (x-a)\,(\sqrt{b-a} - \sqrt{x-a})\,u''(x)\,dx\,,$$

we have but to do the same considerations and computations as in § 5.3.

In correspondence with (1) we have

$$\varphi_0(x) = \frac{4}{3}(x-a)^{3/2}, \quad \varphi_1(x) = -\frac{4}{3}(b-a)^{1/2}(x-a) + \frac{4}{3}(x-a)^{3/2},$$

$$\varphi_2(x) = \frac{4}{3}(x-a)^{3/2} - 2(b-a)^{1/2}(x-a) + \frac{2}{3}(b-a)^{3/2},$$

whence, writing $\varphi_1(x)$ in the form $\frac{1}{2}[\varphi_0(x) + \varphi_2(x)] + c_{11}(x-a) + c_{12}(b-x)$, we see that

$$c_{11} = -\frac{2}{3}(b-a)^{1/2}, \qquad c_{12} = -\frac{1}{3}(b-a)^{1/2}.$$

We then have

$$\varphi_0(x;\lambda,\mu) = \int_a^x \frac{x-\xi}{\sqrt{\lambda\xi + (1-\lambda)\mu - a}}\, d\xi,$$

$$\varphi_2(x;\lambda,\mu) = -\int_x^b \frac{x-\xi}{\sqrt{\lambda\xi + (1-\lambda)\mu - a}}\, d\xi,$$

$$\varphi_1(x;\lambda,\mu) = \frac{1}{2}[\varphi_0(x;\lambda,\mu) + \varphi_2(x;\lambda,\mu)] + c_{11}(\lambda,\mu)(x-a) + c_{12}(\lambda,\mu)(b-x)$$

and it is convenient to choose $c_{11}(\lambda,\mu)$, $c_{12}(\lambda,\mu)$ in such a way that it be $\varphi_1(a;\lambda,\mu) = \varphi_1(b;\lambda,\mu) = 0$: we find

$$c_{11}(\lambda,\mu) = -\frac{1}{2(b-a)} \int_a^b \frac{b-\xi}{\sqrt{\lambda\xi + (1-\lambda)\mu - a}}\, d\xi,$$

$$c_{12}(\lambda,\mu) = -\frac{1}{2(b-a)} \int_a^b \frac{\xi-a}{\sqrt{\lambda\xi + (1-\lambda)\mu - a}}\, d\xi,$$

whence there results

$$\varphi_1(x;\lambda,\mu) = -\int_a^x \frac{(b-x)(\xi-a)}{b-a} \frac{d\xi}{\sqrt{\lambda\xi + (1-\lambda)\mu - a}}$$

$$-\int_x^b \frac{(x-a)(b-\xi)}{b-a} \frac{d\xi}{\sqrt{\lambda\xi + (1-\lambda)\mu - a}}.$$

Then, we may write the following elementary quadrature formula:

$$\int_a^b \frac{u(x)\,dx}{\sqrt{\lambda x + (1-\lambda)\mu - a}} = A_{01}(\lambda, \mu)\, u(a) + A_{02}(\lambda, \mu)\, u(b) + R(u; \lambda, \mu)\,, \tag{2}$$

with

$$\left.\begin{aligned} A_{01}(\lambda, \mu) &= -\varphi_1'(a) = \int_a^b \frac{b-\xi}{b-a} \frac{d\xi}{\sqrt{\lambda \xi + (1-\lambda)\mu - a}}\,,\\ A_{02}(\lambda, \mu) &= \varphi_1'(b) = \int_a^b \frac{\xi - a}{b-a} \frac{d\xi}{\sqrt{\lambda \xi + (1-\lambda)\mu - a}}\,,\\ R(u; \lambda, \mu) &= \int_a^b \varphi_1(x; \lambda, \mu)\, u''(x)\, dx\,. \end{aligned}\right\} \tag{3}$$

It is soon seen that the hypothesis (5.2.17) is satisfied [that is $c_{11}(1, \mu) = c_{11}$, $c_{12}(1, \mu) = c_{12}$] and also (5.2.21) since we have

$$|c_{11}(\lambda, \mu)|,\ |c_{12}(\lambda, \mu)| < \frac{1}{2} \int_a^b \frac{d\xi}{\sqrt{\lambda \xi + (1-\lambda)\mu - a}}\,,$$

$$\begin{aligned} &\lim_{\lambda \to 0} \frac{\lambda}{2} \int_a^b \frac{d\xi}{\sqrt{\lambda \xi + (1-\lambda)\mu - a}}\\ &\quad = \lim_{\lambda \to 0} \left[\sqrt{\lambda b + (1-\lambda)\mu - a} - \sqrt{\lambda a + (1-\lambda)\mu - a}\right] = 0 \end{aligned}$$

uniformly for $\mu \in [a, b]$.

Starting from (2), we may now get the generalized formula:

$$\int_a^b \frac{u(x)\,dx}{\sqrt{x-a}} = \sum_{j=1}^{\nu} \lambda_j \left[A_{01}(\lambda_j, \mu_j)\, u(\alpha_{j-1}) + A_{02}(\lambda_j, \mu_j)\, u(\alpha_j)\right] + \varrho(u)$$ [1])

which can also be written

$$\begin{aligned} \int_a^b \frac{u(x)\,dx}{\sqrt{x-a}} &= \lambda_1 A_{01}(\lambda_1, \mu_1)\, u(a) + \sum_{j=1}^{\nu-1} [\lambda_{j+1} A_{01}(\lambda_{j+1}, \mu_{j+1})\\ &\quad + \lambda_j A_{02}(\lambda_j, \mu_j)]\, u(\alpha_j) + \lambda_\nu A_{02}(\lambda_\nu, \mu_\nu)\, u(b) + \varrho(u)\,. \end{aligned} \tag{4}$$

Let us now try to simplify (4) by choosing suitable points α_j. From formulae (3), putting $\lambda = \lambda_j$, $\mu = \mu_j$ and operating the substitution $\lambda_j \xi + (1 - \lambda_j)$

[1]) Here the symbols α_j, δ_j, λ_j, μ_j have the same meanings that they had in § 5.2.

$\times\ \mu_j - a = t^2$, we obtain

$$\left.\begin{aligned} \lambda_j A_{01}(\lambda_j, \mu_j) &= \frac{2}{\delta_j} \int\limits_{\sqrt{\alpha_{j-1}-a}}^{\sqrt{\alpha_j - a}} (\alpha_j - a - t^2)\, dt\,, \\ \lambda_j A_{02}(\lambda_j, \mu_j) &= \frac{2}{\delta_j} \int\limits_{\sqrt{\alpha_{j-1}-a}}^{\sqrt{\alpha_j - a}} (t^2 - \alpha_{j-1} + a)\, dt\,. \end{aligned}\right\} \tag{5}$$

It is convenient to choose α_j in such a way that the points $\sqrt{\alpha_j - a}$ divide the interval $[0, \sqrt{b - a}]$ in ν equal parts, namely to choose $\alpha_j = a + \frac{j^2}{\nu^2}(b - a)$ and therefore $\delta_j = \frac{2j - 1}{\nu^2}(b - a)$. With such a choice, formulae (5) give

$$\lambda_j A_{01}(\lambda_j, \mu_j) = \frac{2}{3\nu} \frac{3j - 1}{2j - 1} \sqrt{b - a}\,, \qquad \lambda_j A_{02}(\lambda_j, \mu_j) = \frac{2}{3\nu} \frac{3j - 2}{2j - 1} \sqrt{b - a}\,,$$

so that (4) becomes

$$\int\limits_a^b \frac{u(x)\, dx}{\sqrt{x - a}} = \frac{2}{3\nu} \sqrt{b - a} \left\{ 2\, u(a) + 4 \sum_{j=1}^{\nu-1} \frac{3j^2 - 1}{4j^2 - 1} \right.$$

$$\left. \times\ u\left[a + \frac{j^2}{\nu^2}(b - a)\right] + \frac{3\nu - 2}{2\nu - 1}\, u(b) \right\} + \varrho(u)\,, \tag{6}$$

where, on the basis of theorem 5.2.III, we have $\varrho(u) = o\left(\frac{1}{\nu}\right)$ (for $\nu \to \infty$).

BIBLIOGRAPHY

[1] ACHIESER, N. I. 1967, *Vorlesungen über Approximations Theorie* (Akademie Verlag, Berlin).

[1] BEREZIN, I. S. 1965 (with N. P. ZHIDKOV) *Computing methods*, I – II (Pergamon Press, London).

[1] BOUZITAT, J. 1949, *Sur l'integration numérique approchée par la méthode de Gauss généralisée et sur une extension de cette méthode* (C. R. Acad. Sc. Paris, 229).

[1] CAPUANO, R. See SALZER–ZUCKER–CAPUANO [1].

[1] DAVIS, P. J. 1956 (with P. RABINOWITZ) *Abscissas and weights for gaussian quadratures of high order.* (J. Res. Nat. Bur. Standards, 56).

[2] 1958 (with P. RABINOWITZ) *Abscissas and weights for gaussian quadratures of high order: values for* $n = 64$, 80 *and* 96 (J. Res. Nat. Bur. Standards, 60).

[1] EMDE, F. see JAHNKE–EMDE–LOSCH [1].

[1] GHIZZETTI, A. 1954, *Sulle formule di quadratura* (Rendiconti del Seminario Matematico e Fisico di Milano, XXVI).

[2] 1956, *Sulla convergenza dei procedimenti di calcolo degli integrali definiti forniti dalle formule di quadratura* (Rendiconti del Seminario Matematico dell'Università di Padova, XXVI).

[3] 1962, *Sulle formule di quadrature relative ad intervalli illimitati* (Rendiconti dell'Accademia dei Lincei, XXXII).

[4] 1962, *Les formules de quadrature sur les intervalles infinis* (Annales de la Faculté des Sciences de l'Université de Clermont, II).

[5] 1968, *Procedure for constructing quadrature formulae on infinite intervals* (Numerische Mathematik, 12).

[6] 1967 (with A. OSSICINI) *Su un nuovo tipo di sviluppo di una funzione in serie di polinomi* (Rendiconti dell'Accademia dei Lincei, XLIII).

[1] GORI, L. 1960, *Una generalizzazione della formula di Lagrange-Hermite* (Ricerche di Matematica, IX).

[1] GRENWOOD, R. E. 1948 (with J. J. MILLER) *Zeros of the Hermite polynomials and weights for Gauss mechanical quadrature formula* (Bull. Am. Math. Soc., 54).

[1] HILDEBRAND, F. B. 1967, *Analisi Numerica* (Casa Editrice Ambrosiana, Milano).

[1] HOUSEHOLDER, A. S. 1953, *Principles of numerical analysis* (McGraw-Hill Book Company Inc., New York).

[1] JACKSON, D. 1921, *The general theory of approximation by polynomials and trigonometric sums* (Bull. Am. Math. Soc., 27).

[2] 1930, *The theory of approximation*, vol. I, II (Am. Math. Soc. Colloquium Publ., New York).

[1] JAHNKE, E. 1960 (with F. EMDE and F. LOSCH) *Tables of higher functions* (McGraw-Hill Book Company Inc., New York).

[1] KARLIN, S. J. 1966 (with W. J. STUDDEN) *Tchebycheff systems with applications in analysis and statistics* (Interscience Publishers, New York).

[1] KNESCHKE, A. 1949, *Theorie der genäherten Quadratur* (J. reine angew. Math., 187).

[1] KOPAL, Z. 1962, *Numerical analysis* (John Wiley & Sons, Inc., New York).

[1] KRILOV, V. I. 1962, *Approximate calculation of integrals* (MacMillan Company, New York).

[1] LOSCH, F. See JAHNKE–EMDE–LOSCH [1].
[1] MANSION, P. 1914, *Théoreme général de Peano sur le reste dans les formules de quadrature* (Mathesis, 34).
[1] MILLER, J. J. See GREENWOOD-MILLER [1].
[1] MINEUR, H. 1952, *Techniques de calcul numérique. Note de J. Bouzitat* (Libraire Polytechnique, Paris).
[1] MISES, R. (von) 1935, *Ueber allgemeine Quadraturformeln* (J. reine angew. Math., 174).
[1] MORELLI, A. 1967–68, *Formula di quadratura con valori della funzione e delle sue derivate anche in punti fuori dell'intervallo d'integrazione* (Atti dell'Accademia delle Scienze di Torino, CII).
[1] NILSEN, K. L. 1964, *Methods in numerical analysis* (McMillan Company, New York).
[1] OBRESCHKOFF, N. 1940, *Neue Quadraturformeln* (Abh. Preuss. Akad. Wiss., 4).
[1] OSSICINI, A. 1966, *Costruzione di formule di quadratura di tipo gaussiano* (Annali di Matematica Pura ed Applicata, LXXII).
[2] 1967, *Sulle costanti di Christoffel della formula di quadratura di Gauss-Jacobi* (Istituto Lombardo di Scienze e Lettere, 101).
[3] 1968, *Le funzioni di influenza nel problema di Gauss sulle formule di quadratura* (Le Matematiche, XXIII).
[4] See GHIZZETTI–OSSICINI [6].
[1] PEANO, G. 1913, *Resto nelle formule di quadratura espresso con un integrale definito* (Rendiconti della Reale Accademia dei Lincei, XXII).
[2] 1914, *Residuo in formulas de quadratura* (Mathesis, 34).
[1] PICONE, M. 1951, *Vedute generali sull'interpolazione e qualche loro consequenza* (Annali Sc. Norm. Sup. Pisa V).
[1] POPOVICIU, T. 1955, *Asupra unei generalizări a formulei de integrare numerica a lui Gauss* (Academia R. P. R. Filiale Iasi Studii Si Çercetari Stientifice, VI).
[2] 1964, *La simplicité du reste dans certaines formules de quadrature* (Mathematica, 6).
[1] RABINOWITZ, P. 1959 (with G. WEISS) *Tables of abscissas and weights for numerical evaluation of integrals of the forme* $\int_0^{+\infty} e^{-x} x^n f(x)\,dx$ (Math. Tables., Aids Comput., 13).
[2] See DAVIS–RABINOWITZ [1].
[3] See DAVIS–RABINOWITZ [2].
[1] RADON, J. 1935, *Restausdrücke bei Interpolations und Quadraturformeln durch bestimmte Integrale* (Monatsh. Math., 42).
[1] REBOLIA, L. 1966, *Formule di quadratura di tipo gaussiano con valori delle derivate dell'integrando* (Calcolo, 3).
[1] ROGHI, G. 1967, *Sul resto delle formule di quadratura di tipo gaussiano* (Le Matematiche, XXII).
[1] ROSATI, F. 1968, *Problemi di Gauss e di Tchebychef relativi a formule di quadratura esatte per polinomi trigonometrici* (Le Matematiche, XXIII).
[1] SALZER, H. E. 1952 (With R. ZUCHER — R. CAPUANO) *Table of the zeros and weight factors of the first twenty Hermite polynomials* (J. Res. Nat. Bur. Standards, 48).
[2] 1954 (with R. ZUCHER) *Table of the zeros and weight factors of the first fifteen Laguerre polynomials* (National Bureau of Standards).
[1] SECREST, D. See STROUD–SECREST [1].
[1] SINGER, J. 1964, *Elements of numerical analysis* (Academic Press, London).
[1] STEFFENSEN, J. F. 1965, *Interpolation* (Chelsea Publishing Company, New York).
[1] STROUD, A. H. 1966 (with D. SECREST) *Gaussian quadrature formulas* (Prentice-Hall, Inc., Englewood Cliffs, N.J.).
[1] STUDDEN, W. J. See KARLIN–STUDDEN [1].
[1] SUPPA, R. 1963, *Sulle condizioni affinchè un operatore differenziale lineare sia autoaggiunto* (Pubblicazione I. N. A. C., Quaderno n. 3).

Bei Fragen zur Produktsicherheit wenden Sie sich bitte an:
If you have any questions regarding product safety,
please contact:

Walter de Gruyter GmbH
Genthiner Straße 13
10785 Berlin
productsafety@degruyterbrill.com